Patrick Moore's
Practical Astronomy Series

Springer
London
Berlin
Heidelberg
New York
Barcelona
Hong Kong
Milan
Paris
Singapore
Tokyo

Other titles in this series

The Observational Amateur Astronomer
Patrick Moore (Ed.)

Telescopes and Techniques
Chris Kitchin

The Art and Science of CCD Astronomy
David Ratledge (Ed.)

The Observer's Year
Patrick Moore

Seeing Stars
Chris Kitchin and Robert W. Forrest

Photo-guide to the Constellations
Chris Kitchin

The Sun in Eclipse
Michael Maunder and Patrick Moore

Software and Data for Practical Astronomers
David Ratledge

Amateur Telescope Making
Stephen F. Tonkin

Observing Meteors, Comets, Supernovae
and other Transient Phenomena
Neil Bone

Astronomical Equipment for Amateurs
Martin Mobberley

Transit: When Planets Cross the Sun
Michael Maunder and Patrick Moore

Practical Astrophotography
Jeffrey R. Charles

Observing the Moon
Peter T. Wlasuk

Deep-Sky Observing
Steven R. Coe

AstroFAQs
Stephen F. Tonkin

The Deep-Sky Observer's Year
Grant Privett and Paul Parsons

Field Guide to the Deep Sky Objects
Mike Inglis

Choosing and Using a Schmidt–Cassegrain Telescope
Rod Mollise

Astronomy with Small Telescopes
Stephen F. Tonkin (Ed.)

Solar Observing Techniques
Chris Kitchin

Observing the Planets
Peter T. Wlasuk

Light Pollution
Bob Mizon

Using the Meade ETX
Mike Weasner

Practical Amateur Spectroscopy
Stephen F. Tonkin (Ed.)

More Small
Astronomical
Observatories

Patrick Moore (Ed.)

With 203 Figures

Springer

Cover illustration includes Ken Dauzat Observatory, described in Chapter 8. Observatory image courtesy of Ken Dauzat.

British Library Cataloguing in Publication Data
More small astronomical observatories. – (Patrick Moore's practical
    astronomy series)
    1. Astronomical observatories – Design and construction –
    Amateurs' manuals 2. Astronomical observatories – Great Britain
    I. Moore, Patrick, 1923–
    522.1
Additional material to this book can be downloaded from http://extras.springer.com.
ISBN 1852335726

Library of Congress Cataloging-in-Publication Data
A catalog record for this book is available from the Library of
Congress

Patrick Moore's Practical Astronomy Series ISSN 1617-7185
ISBN 1-85233-572-6 Springer-Verlag London Berlin Heidelberg
a member of BertelsmannSpringer Science+Business Media GmbH
http://www.springer.co.uk

Typeset by EXPO Holdings, Malaysia
Printed and bound at the Cromwell Press, Trowbridge, Wiltshire
58/3830-543210 Printed on acid-free paper SPIN 10770924

Preface

In *Small Astronomical Observatories* in this series, the accounts of small observatories – almost entirely amateur – caused a great deal of interest. The descriptions were very useful to many people, and many new observatories were set up as a direct result.

Therefore the time seemed ripe for a second collection, and this is presented here. There are observatories of many kinds. Some, such as that at Long Crendon, are ambitious and enable work of top professional standard to be carried out; others are much simpler – even portable. But all have their strong points, and the would-be observatory builder will find that there is ample guidance here.

Small Astronomical Observatories is out of print, but there is still demand for it. Therefore, we have put it on the CD that is to be found at the back of this new book.

Go to it – and may you all have clear skies!

Patrick Moore
January 2002

Contents

1 Garage and Garden Observatory 1
   Peter Paice

2 A Portable Observatory. 19
   Rob Johnson

3 A User-Friendly Run-Off Shed
   for a 12-inch LX200. 29
   Martin Mobberley

4 Darklight Observatory, Eddyville,
   Kentucky, USA. 41
   Chris Anderson

5 Turner Observatory. 55
   Bob Turner

6 A Simple Rotating Observatory
   in Nottingham, England 63
   Alan W. Heath

7 St Margaret's Observatory 69
   Paul Andrew

8 Ken Dauzat Observatory. 81
   Ken Dauzat

9 A Lancashire Observatory – Part II. 93
   David Ratledge

10 Arcturus Observatory 107
   Paul Gitto

11 Osmundstö Observatory: A Garage
   Observatory for CCD Imaging Located
   at the Shoreline of Southern Norway 121
   Alf Jacob Nilsen

12 Huntington Observatory, York 153
   Mike Brown

13 Ptolemy's Café. 163
   Bill Arnett

14 The Construction of Starbase Two 175
   Paul Zelichowski

15 A Domestic Solar Observatory 189
   George Kolovos

16 Coddenham Observatory in
   Suffolk, England 203
   Tom Boles

17 Building the Crendon Observatory 213
   Gordon Rogers

18 The Marina Towers Observatory, Swansea. . . . 225
   S.J. Wainwright

   Contributors 239

   About the CD-ROM 248

Chapter 1
Garage and Garden Observatory

Peter Paice

Introduction

Observatory is perhaps too grand a title – an observing site would be more accurate! Belfast in Northern Ireland has a climatic regime dominated by Atlantic depressions with attendant cloud cover, so day- or night-time observing is a little frustrating at times.

My house on the outskirts of the City suffers the consequent light pollution levels from several sodium street lamps in close proximity to the south which preclude any deep-sky observing or imaging. However I can transport my telescopes and mount some 3 km (2 miles) into the darker countryside. My latent interest in practical astronomy started when I was stimulated by the excellent coverage of the subject on television. I combined my very long-established hobby of photography, including processing and printing, with my practical skills in electronics, optics and machine-shop practice. I had an old second-hand 6-inch Newtonian with a home-made single-axis quartz drive and I was a convert!

Retirement provided more time for my hobbies but consequent restrictions on disposable income to lavish on them. Consequently my philosophy has been to infrequently purchase ready-made equipment but to adapt or modify my existing stock of photographic items. It would seem inappropriate to purchase very expensive top-of-the-range telescopes and imaging devices only to discover that due to restricted "seeing" conditions they were rarely used. The following account will indicate the development of my equipment and observational methods over some six years.

The Garage Observatory

My initial purchase was a Vixen VC 200L Visac (sixth-order aspherical catadioptric) 1800 mm focal length, f/9

Figure 1.1. The Vixen Visac catadioptric telescope with the Astrovid 2000 and MX5C attached to flip-mirror box. The black-and-white monitor and associated cabling is also shown.

telescope, and a Vixen GP-DX dual-axis equatorial mount fitted to a sturdy Orion Optics heavy-duty tripod. The site faces south. Some conventional film camera astrophotography followed. Then a chance viewing of a TV programme showing a famous science fiction writer viewing real-time lunar images indoors using a video-camera coupled to his telescope prompted me to purchase a black-and-white Astrovid 2000 video-camera with 520 lines-per-inch resolution. Images (1 volt, peak-to-peak) were initially examined on a 700 line-per-inch monitor and then fed from the garage upstairs to a TV and video-recorder. Still images were acquired by a video-card in my PC, saved to file, perhaps enhanced and printed.

Purchase of a CCD (Starlight Xpress MX5C) followed, together with a flip-mirror box. Could this flip-mirror, the dual axes and the focus be controlled remotely? The CCD supplier questioned my intention of sending digital images some 12 m (40 feet) to the PC running the imaging software.

Figure 1.2. The focus control (A), the servo-motor (B) for flipping the mirror, and the CCD camera.

The Astrovid 2000 mounted in this mode and sending real-time video images functions excellently also as a guider for the CCD camera. If the celestial object is aligned with an X marked on the TV screen in my study then the same object is usually within the field of the CCD (see Figure 1.3).

Remote Control of Right Ascension and Declination

This was effected by fitting suitable male in-line plugs on coaxial extension cables running from the garage wall to female sockets in a wall box near the PC desk upstairs in my study. The celestial object is located on the black-and-white garage monitor using the Astrovid camera and the right ascension (RA) and declination hand-controller. The hand-controller is unplugged from the female sockets on the telescope mount, the extension cables plugged in and a fast passage is made to my study and the RA and dec. cables plugged into the wall sockets! Usually the

Figure 1.3. My PC running the CCD imaging software and the Astrovid 2000 image (Saturn) in the guiding mode.

planet image has not disappeared from the TV screen; a tweak on the RA button quickly centres the object. Sounds very complicated but with practice works well and saves purchasing another hand-controller. However there is duplication of rechargeable sealed lead acid packs for powering the controller; these are preferable to mains adapters, giving a more stabilized voltage. Remarkably there was no signal drop-out over the 12 m length of coaxial cable. The remote control of the focus and flip-mirror required a little ingenuity and considerable trial and error!

Remote Control of Focus

Reference to Figure 1.1 and A in Figure 1.2 shows the motor drive of a 3 V electric screwdriver coupled via a step-down 3:1 ratio radio tuning dial connector and a further step-down ratio of 10:1 provided by a right-angled worm and cog block. This gave a final rotational speed of 2 rpm. Coupling to the telescope was effected by removing a focusing knob and linking the shafts by means of a 10 mm brass compression coupling and a slotted clutch. Drive power comes via a Bulgin three-pin miniature connector, the female socket being attached to the drill by glass fibre resin and supplied by a plug on a 3 V power lead from the study control box, see Figure 1.4. The control box contains rechargeable NiCad cells removed from the electric drill body and a short power lead to check out the focus drive at the telescope before moving the box to the remote study.

Remote Control of the Flip-Mirror

Figure 1.4 shows an upper three-position toggle switch for "flip" or "focus" selection and a lower biased toggle switch, which controls the flip-mirror. In the "U" position the Astrovid video-camera is selected; in the "L" the CCD camera is selected. Reference to Figure 1.2 shows a close-up view of the 3 V model car steering servo (B) and the small lever connection to the mirror fulcrum axle.

All the "remotes" worked well and it made quite a change after acquisition of the target image to work in the warm comfort of my study.

Figure 1.4. The control box with a short 3 V power lead to check focus, and two jack sockets. The two toggle switches allow selection of control modes.

Camera Telephoto Lenses as Imaging Tools

My next experiments involved using several good-quality telephoto lenses purchased to fit my 35 mm Olympus and 6 × 4.5 cm Mamiya cameras as "telescopes". Luckily I posses many extension tubes for these cameras and by careful use of glass fibre resin, aluminium tubes and lathe skills it was possible to make adapters to couple such combinations as a 400 mm Mamiya to the MX5C CCD camera; a 400 mm Sigma or the 400 mm Mamiya linked to the Astrovid 2000 video-camera or a 400 mm Tamron linked to the CCD. Two such combinations are shown in Figure 1.5 (a and b). The Astrovid video-camera has a "C" thread and converters are available for several major camera makes.

Introduction of Digital Cameras and Afocal Projection

Purchase some four years ago of a Fugi MX 700 with 1.5 megapixels for "normal" photography and having a "smart media" card seemed ideal for experimentation with astrophotography. Knowledge of afocal projection led me to try and image a distant chimney using the 7 × 50 mm finderscope of my Vixen Visac telescope. It worked! Fortunately this camera was a non-zoom type and possesses a self-timer but unfortunately the longest exposure was about 3 seconds. I set about making a "holster" to carry and connect the camera to the standard 37.1 mm diameter long eye relief eyepieces. My experience with cutting, lathe turning and cementing acrylic-type plastics for underwater camera cases came in handy for making a Mk I holster (see Figure 1.6). Short lengths of various plastic vacuum cleaner and wash-basin tubes were found to have a second useful life as they allowed a slide-fit with the eyepieces.

The Fugi MX 700 in the holster can be used for afocal projection with all my telescopes. I made my excellent Mamiya 200 mm telephoto lens and custom ×2 converter into a telescope by fitting a 31.7 mm diagonal prism inside a machined Mamiya-fit extension tube (glass fibre again!) with provision for any of my eyepieces (illustrated in Figure 1.7). This set-up for sunspot imaging takes just ten minutes from bringing the tripod out of my garage and aligning the "pods" with drilled holes in the concrete driveway. The "up-and-over" garage door is partially lowered as a sunscreen. It may seem a crude method but the alignment and the drives keep the solar image centred for several hours! The same situation pertains for lunar imaging. When required, the Vixen catadioptric telescope can also accommodate the digital camera set-up for higher magnification.

The Fugi MX 700 with the maximum exposure of some 3 seconds limited it to solar and planetary imaging. The saved images on smart media cards were easily transferred to my PC. However with the "third generation" of digital cameras comes more pixels (not in itself giving higher resolution), and larger frame images requiring image transfer cards but luckily most have direct USB transfer to a PC. All have a zoom facility, partly optical, partly digital. After due consideration of prices, and specification, an Olympus Camedia 2040Z (2.1 megapixels) was purchased. It had an all-important manual exposure of up to 10 seconds. The zoom capability is particularly useful for solar prominence imaging as it quickly gives increased magnification without eyepiece changing! Next came the task of making a custom-fit holster or

Figure 1.5a. A Sigma 400 mm lens coupled to the Astrovid 2000 video-camera and a Mamiya 400 mm lens and MX5C CCD.

carrier. The usual 12.5 mm ($\frac{1}{2}$-inch) acrylic plastic sheet was chosen for the ease of cutting, machining and cementing; plus the choice from an assemblage of plastic wash-basin outlet pipes and vacuum cleaner tubes! Thus a new-style custom fit camera carrier for the Olympus 2040Z digital camera came into existence. The camera is retained in the carrier by a $\frac{1}{4}$ inch Whitworth bolt of the correct length having a large knurled tightening knob (see Figure 1.8).

Many of the digital cameras with zoom capability have the convex curvature of the front lens slightly protruding from the lens mounting. Consequently when constructing the eyepiece tube of the plastic camera carrier there must be a stop-ring cemented to the inside of the tube wall to prevent the eyepiece lens

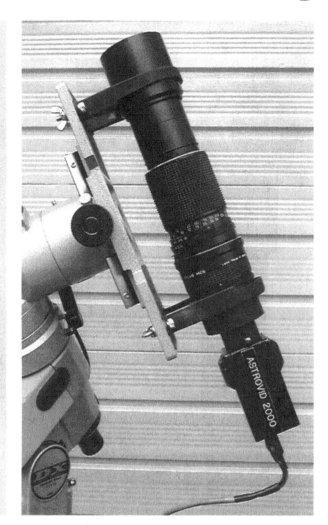

Figure 1.5b. A Mamiya 200 mm plus a ×2 converter coupled to the Astrovid video-camera.

(often with a metal seating) scratching the camera lens! For my carrier I prepared a 2 mm thick plastic stop-ring of the tube material, removed a 4 mm section of the ring circumference so that on slight compression it would just start to slip up the carrier tube (like fitting an engine piston ring!). The zoom lens control on the camera was activated and the protruding zoom mount carefully used to push the split ring to the correct position, the camera gently removed and the split ring cemented using an "instant" glue. It works extremely well, so camera and eyepiece lenses can never touch.

Figure 1.6a. The picture shows the Mk I holster to accommodate the Fugi MX 700 digital camera.

The Garden Observatory

Being moderately free from the glare of street lamps, the garden observatory utilizes the portability and flexibility of the new digital camera and carrier. A Celestron 6-inch, f/8 (tube only) refractor was purchased and mounted on the Vixen GP mount and the very sturdy Orion Optics tripod. A suitable concrete base was constructed to a depth of 45 cm ($1\frac{1}{2}$ feet). Polar alignment was relatively easy as there is an almost unobstructed view of Polaris, and the Vixen mount has an internal polar alignment telescope. Tripod location holes were drilled into the concrete for speedy set-up. The garden site also allows good visual access to the north-western and south-eastern sky but trees and the greenhouse partly obstruct the lower northern sky (see Figure 1.9).

Figure 1.6b. The picture shows the digital camera in place. Brass M4 setscrews centre and grip the eyepieces.

Both the Vixen catadioptric telescope and the Celestron refractor are used at the garden site with either the Fugi MX 700 or Olympus 2040 zoom digital cameras (see Figure 1.10). Images are stored on internal 32 Mb or 64 Mb removable smart media cards. The darker garden site is obviously preferred for planetary imaging and attempts at deep-sky imaging within the restraints of a 10-second exposure. Solar imaging of sunspots and prominences is usually conducted at the garage site using the modified Mamiya telescope or the Celestron refractor.

An Experimental Coronagraph Joins the Garage Observatory

The stimulus to make a simple coronagraph came after seeing the solar coronal images from the SOHO

Figure 1.7. The modified 400 mm Mamiya telephoto lens. This afocal projection into the Fugi MX 700 digital camera gives excellent solar or lunar images.

Figure 1.8a. The custom-fit camera carrier.

Figure 1.8b. The Olympus 2040Z digital camera in place on the carrier.

Figure 1.9. The 6-inch, f/8 Celestron refractor on the Vixen GP equatorial mount and sturdy Orion Optics tripod. Also shown is a Thousand Oaks 1.5 Å solar prominence filter set in position and afocal coupling to the Olympus 2040Z digital camera.

Figure 1.10. The lower end of the Celestron 6-inch refractor with the tuneable part of the 1.5 Å prominence filter attached via a diagonal prism and a 40 mm eyepiece afocally coupled to the Olympus 2040Z.

Figure 1.11a. The occulting disk mounted on the filter in the ring.

satellite, available at the NASA Web site. Historically one of the first successful coronagraphs was made by Bernard Lyot in France in about 1932. Earlier attempts to create an artificial total eclipse of the Sun within the telescope system had failed. The NASA coronagraph image suggested that by inserting an occulting disk just above the film in a camera attached to my home-constructed 5-inch, f/10 refractor might just work. Would the occulting disk become red-hot? Luckily I managed to make a mirror lock-up catch within my lightweight Vixen 35 mm astro-camera (Pentax fitting) so the occulting disk could be lowered down to just above the shutter blind. Several attempts at designing and making occulting disks took place as I did not want to make an occulting disk that itself made the solar corona by reflecting light rays from the disk edge or support system. The occulting disk was a turned copper disk with a small central boss turned on my lathe 2 mm wider in diameter than the solar image on the film surface. It was silver soldered to a stainless steel 17-gauge serum needle, the needle shaft having been cut to the correct length.

Figure 1.11b. The occulting disk in position at the prime focus of the refractor.

The final design required the disk to be cemented on a neutral glass photographic filter residing in a turned circular recess of a Pentax 10 mm lens extension ring (see Figure 1.11).

Figure 1.12a.
Alignment of the occulting disk.

With the equatorial mount drives running, the image of the solar disk is projected on to a white card and the drives incrementally adjusted to centre the solar image precisely into the occulting disk. Next the camera is checked for film advance, mirror locked up and in self-timer mode. Using a cable release, the exposure is made. Unfortunately this set-up has to be repeated for each exposure because advancing the film usually affects the accuracy of the alignment (see Figure 1.12).

Finally

My method of digital astro-imaging may be frowned upon by some astrophotographers, but for myself it has been a challenge to develop the technology as far as my limited finance allows. The ability to keep or discard images, save to a PC file, colour print or send to distant Web sites is technically exciting, but perhaps most important for me is the portability of the telescope to dark sites. Many of my astro-images using this method have been published or are on astro Web sites. Imaging

Figure 1.12b. Typical solar corona produced by the coronagraph (from a 200 ASA colour negative).

via afocal projection to digital cameras has limitations, but for deep-sky imaging the specialist CCD and use of sensitized film must continue.

Clear skies!

Chapter 2

A Portable Observatory

Rob Johnson

Like many amateur astronomers, I have always wanted a large permanently sited telescope in an observatory, but I have always been limited by having a small garden. A traditional domed observatory was out of the question and a roll-off roof or run-off shed type of observatory wouldn't save much on space either. The answer was to permanently site the mount and make the rest of the "observatory" portable. The ideas for my portable observatory started to develop with my first telescopes.

A Permanent Mount

My first serious telescope was a 150-mm (6-inch) reflector mounted on a home-made German equatorial. I wanted the mount to be polar aligned as accurately as possible to take full advantage of the equatorial, but alignment every observing session was not an option. These were the days long before today's mounts with built-in polar finding scopes: my pipe flange mount would have been too difficult and time-consuming to set up each session. My solution was to permanently mount the RA axis on top of a pillar that was concreted into the ground. The rest of the mount was then slotted into the RA axis and the telescope bolted on when I wanted to observe. The RA axis was nothing more sophisticated than a 25-mm (1-inch) diameter steel pipe and so didn't need much in the way of protection from the weather apart from a bit of grease. The telescope and the rest of the mount were stored indoors. This system worked very well for several years until aperture fever overcame me and I hatched plans for a 360-mm (14-inch) reflector.

The demands on my new telescope would be much greater. As my interest had grown in astronomy I became increasingly interested in astrophotography, so the new telescope would have to be up to the rigours of long-exposure photography through the telescope. I was lucky enough to come by a very well-engineered German equatorial mount for sale through my local club. The mount had been cast in aluminium with large bearings on each axis. The mount was ideal for a 14-inch (Figure 2.1) but in no way could the 60 kg (132 lbs) mount be lifted out each observing session.

The Mount Cover

I decided to permanently site the whole mount and build a lightweight cover to protect it from the weather. The cover would have to be quick and easy enough to

Figure 2.1. The 14-inch reflector and the German equatorial mount.

remove but protect the mount, its RA worm wheel drive, motor and electronics from the worst of the British winter.

I opted to construct the cover using a wooden frame made out of lengths of 25-mm (1-inch) square hardwood and built to the shape of the mount in its storage position (Figure 2.2). Building this shape would be more complex than a simple cube but would lessen wind resistance and allow the rain to run off the slopes. All of the joints were glued and screwed together, with extra strength provided at the angled joints by plywood blocks.

Another of my objectives for the cover was that it should look as aesthetically pleasing as possible in the garden. For this reason, and also one of cost, I decided to cover the wooden frame in polythene sheet. After a few months this idea hit serious problems as the polythene started to become brittle and crack due to the action of sunlight. A friend of mine, Kevin Johnson, suggested I replace the polythene with "Monoplex" pond liner. This was a heavier gauge polythene sheet which had a reinforcing mesh built-in. This grade was designed to have excellent UV resistance; what's more it was supplied in a very pale green colour, ideal for use in the garden.

Figure 2.2. The telescope mount cover.

The Monoplex sheet was attached to the wooden frame in two pieces: the first was wrapped around the base of the frame and the second piece would form the roof slopes. The sheeting was secured by wrapping fully around the wooden frame and secured inside the frame with small wood-screws and eyelets. The roof sheet was also secured by the same method from the outside, though I also found weatherproof adhesive tape to be a good alternative.

The assembled structure was strong and extremely lightweight, which of course made an ideal kite! To ensure the cover stayed put on windy days, the base of each long side was fitted with a 90 × 130 mm (3.5 × 5.1 inch) aluminium plate which was drilled with a 25-mm (1-inch) hole. The plates would then fit over 10-mm (0.4-inch) diameter plastic bolts cemented into the concrete base. When the cover was lowered over the bolts, large plastic knurled knobs, made by my friend Dave Galvin, were screwed on to keep the cover firmly in place.

The concrete base itself was sunk into the ground about 610 mm (24 inches) at a diameter of 250 mm (10 inches) or so. At ground level the concrete was spread out to form a flat plinth 900 × 1400 mm (35.5 × 55 inches) and 50 mm (2 inches) deep. On top of this a shallow lip was cast in concrete all the way around the plinth with a 40-mm (1.6-inch) cross-section (Figure 2.3). The lip was made so that the cover would fit snugly over it to form a weatherproof seal. Towards one end of the plinth I made a stubby concrete pillar that would support the mount. In keeping with my desire to have things look aesthetically pleasing, I decided to cast the pillar as a conical section. A very large plastic plant-pot provided an ideal mould when turned upside-down! An earlier version was cast as a pyramid. Four rag bolts were sunk into the top whilst the concrete was still wet. These would hold the mount firmly in position. Mains electricity was provided from an underground cable to the garage.

During 15 years of operation the cover remained waterproof and even in the fiercest gales never parted company with the ground. To my surprise, the steel components of the mount, such as the 32-mm ($1\frac{1}{4}$-inch) shafts, always kept fairly free of rust. I put this down to the fact that the cover acted as a mini-greenhouse and so on sunny days the inside would warm up and keep everything very dry.

Figure 2.3. The German equatorial mount on its concrete base. Note the weatherproof concrete lip.

The Telescope Trolley

The telescope tube assembly weighed 44 kg (97 lbs) and so another novel idea was required to transport the telescope to the mount in the easiest manner. Much of the weight of the telescope tube assembly was due to the primary mirror. It is usual these days to make mirrors over 8 inches or so with less than the ideal diameter-to-thickness ratio of 6:1 and support the mirror carefully to avoid flexure and consequent aberrations. My mirror was unusual as it had been made with the full 6:1 ratio and so was nearly 60 mm (2.5 inches) thick. This allowed the mirror to be simply supported but had the downside that the mirror weighed in at 14 kg (31 lbs)! Another problem this raised was that the natural balance point of the tube assembly would be very close to the mirror, making the eyepiece height at the zenith impossibly high. To place the balance point at a convenient location along the tube I had to use small lead counterweights at the top end of the tube. The cradle, mirror cells and 16-gauge aluminium tube added a further 30 kg (80 lbs) to the total weight.

To easily transport this weight to the mount clearly required something on wheels. I have spent most of my life working in laboratories where large gas cylinders were regularly moved around on a simple vertical trolley, or hand-truck, with two wheels at the base. This gave me the idea for my telescope trolley (Figure 2.4). I made the original trolley frame out of heavy steel angle iron and fitted two 175-mm (6.9-inch) diameter wheels at one end of the frame. The wheels came from a scrap-yard and had 50-mm (2-inch) wide hard rubber tyres. This prototype worked fine but the trolley contributed a lot of unnecessary weight, so I replaced the steel frame with 19-mm (0.75 inch) plywood board that provided the required strength but was much more lightweight.

Using the Telescope Trolley

The telescope was stored outdoors in a small brick-built shed, which was originally used to house an outside toilet! Most of my neighbours had long since demol-

Figure 2.4. The 14-inch telescope and telescope trolley.

ished their "out-houses" but mine remained and was ideal to store my telescope. Getting the telescope out to the mount was very simple. The trolley was attached to the telescope with the same two bolts that held the telescope on the mount. Hand-tightening the nuts was sufficient to keep the telescope and trolley together and this was the way the two were stored. The tube was wheeled out to the mount and stopped in a position close to the baseplate, which was held in a vertical position by propping up the Dec shaft with a plank of wood. The trolley was then unbolted, the tube rotated to face the baseplate and with a short vertical lift, mated up to the baseplate bolt holes. Once the nuts were tightened, the plank could be removed and the telescope, now balanced on its mount, swung into action.

During the planning stage my constant worry was that the frequent transportation back and to the mount would mean that the telescope optics would need to be re-aligned regularly. I built the mirror cell and flat holder as robust but as lightweight as possible and this seemed to pay off. Collimation was only required at most every year; even then only slight adjustments were required. The whole system also performed very well for deep-sky photography; I regularly made guided exposures of 1 hour with good star images.

CCD Equipment Box

During the first few years of operation I used the telescope mainly for deep-sky photography using Technical Pan 2415 and filters to overcome the ever-present light pollution at my location near Liverpool, England. When CCDs appeared on the scene they presented an ideal opportunity for my imaging adventures especially to help combat light pollution. I bought my Starlight Xpress CCD in 1996 for £750, which didn't leave me much of a budget for a computer to run the CCD from, so I decided to build my own 486 desktop. I started with a few odds and ends in my attic and purchased a few parts from a computer fair; the whole bill came to only £100. I housed the computer and monitor in a replica of the computer workstation (Figure 2.5) featured in *Sky & Telescope*, March 1994. The workstation was heavy and bulky but could be fairly easily wheeled out to the telescope.

Figure 2.5. The computer workstation with CCD electronics.

My initial experiences with the CCD, as many amateurs have found, were unforgettable. The downside was the loss of my quick and easy telescope set-up. Getting ready for an observing session was now becoming too tedious and time-consuming; even with the workstation there were still too many loose cables and connections to be made for the CCD.

An old laptop bought at a computer fair provided the answer to my problems. I was now able to build a small wooden case out of 13-mm ($\frac{1}{2}$-inch) frame and three-ply plywood that was carefully designed to house all of the CCD components, the laptop computer and various accessories such as a red torch (Figure 2.6). Each of the main pieces of equipment had its own compartment; the space under the laptop housed the drawtube and CCD filter holder. The case lid was used to attach star maps and observing lists, etc. As many of the connections and cables as possible were built into the

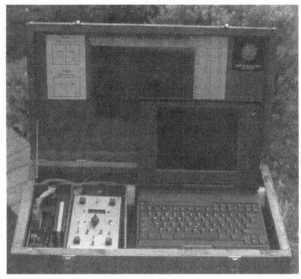

Figure 2.6. Equipment box with laptop PC, CCD electronics and telescope hand-pad controller.

case to simplify setting up. A plug-board inside the box provided mains power so that only one connection to the box was required. The laptop-to-CCD parallel port connection was also semipermanently wired in the case. I chose to build my own case but many amateurs have made use of aluminium photographic equipment cases for this purpose.

When the telescope was set up the equipment case was carried out, connected to the mains supply and placed on a small table. The CCD head could then be connected to the telescope and powered up. An additional advantage with the case was that the CCD could easily be set up and used anywhere with a suitable power supply. On several occasions I have taken the CCD to our club observatory to use on different instruments during star parties.

Conclusions

At first sight my portable observatory may seem like a nightmare, but with some careful thought each part of the set-up can be simplified so that getting ready for an observing session becomes very fast and easy. From start to finish takes just under 10 minutes to set up and be ready for my observing session. In fact it now takes me longer to don all of my winter clothing than it takes to set-up the telescope!

There are also some advantages to having the telescope out in the open. Observatory buildings are notorious for producing turbulent air which helps give rise to poor seeing. With my arrangement I have found that poor seeing is invariably due to the atmosphere and not the telescope or surroundings. Being in the open also allows you to see more of the sky, though it does have the downside that the observer is unprotected from the elements – a small price to pay!

Chapter 3
A User-Friendly Run-Off Shed for a 12-inch LX200

Martin Mobberley

In 1997 I found myself faced with an instrumentation dilemma: technology and anno domini were starting to overtake me! Most amateurs might consider that owning a 36 cm Cassegrain/Newtonian (at my parents' dark-sky site) and a 49 cm Newtonian (at my light-polluted site) would leave me with little need for new equipment. However, I found that I was increasingly in need of an easy-to-use instrument for deep-sky work at the former site. There were a number of reasons for this:

1. The 36 cm telescope was increasingly being used for visual/CCD observation of Jupiter and Saturn and I did not want to disturb the Cassegrain collimation by continually changing to Newtonian mode.

2. Increasing back problems meant I was finding pushing the weighty 36 cm telescope's run-off shed, man-handling the 100 kg (220 lb) tube/counterweight assembly and climbing ladders in the dark, more and more of a battle every clear night: I was starting to dread observing sessions! And so often, just as the telescope was eventually set up, it would cloud over!

3. The proven abilities of the ubiquitous Meade LX200 started to look more and more attractive, even if having the same telescope as thousands of other amateurs seemed a rather sad prospect.

So I decided I would create an observatory maximized for ease-of-use instead of sheer aperture. Once again, a run-off shed would be designed, but this one would be compact and a joy to open and roll back. In addition, it would offer the option of remote observing from indoors.

At a very early stage I decided on a 30 cm LX200 as the best instrument to go for. Many other amateurs in the UK were using this model successfully and it combined a decent aperture with compactness. In addition, using an f/3.3 telecompressor for a focal length of around 1000 mm was ideal for deep-sky/ comet imaging. For example, with an SBIG ST7 CCD camera, I would have a CCD field of view of 24' × 16' and a resolution of just under 2"/pixel. With LX200s capable of slewing to ±5', I should get the target on the CCD chip every time! Once the telescope choice had been made and the dimensions of the instrument obtained, the shed design could proceed.

Preconstruction Considerations

The first decision was how far to mount the observatory from the house. In an ideal world, all observatories would be well above surrounding obstructions like trees and houses; however, my prime concern was what was the longest length of cable I could expect CCD images to travel down before degradation took place; in the end, after consultation, a distance of 13 metres (42 feet) from the east wall of the house was settled on. This meant that the western aspect, as seen from the telescope, has a significant obstruction in the form of the house roof, but there are few other obstructions from the east through to the south-west. I could find no suppliers of long cables for the SBIG ST7 CCD camera (despite the camera being so well suited to remote observing). However, Terry Platt of Starlight Xpress CCDs kindly made me a custom parallel port lead of just over 15 metres (50 feet) in length, despite the fact that, for once, the camera was not one of his! As I already had the ST7, I was able to test the CCD with the long lead attached before the lead was irreversibly buried below the garden, along with the simple RS 232 lead for controlling the telescope. The images using the long lead were clean, with no patterning or undue noise – 15 metres was not a problem, phew! Undoubtedly the high quality of the well-screened lead supplied by Terry was a major factor here.

The mains power cable also travels under the garden and up the side of the plinth. The power comes from a surge-protected earth leakage breaker supply in the house study; this ensures a smooth mains (we live in the country where the mains is rough!) and protects against electrocution (it gets damp outdoors at night). Nine power sockets were eventually screwed to the plinth and an emergency battery back-up supply is also available.

The Plinth

As with my larger telescopes, I decided to utilize a concrete drainage pipe for the plinth for the LX200. A 1 metre long pipe with, crucially, an internal lip on which a base plate can sit, was acquired. Figure 3.1 shows the plinth hole being dug out and Figure 3.2 shows the plinth in position, concreted firmly into the lawn. Metal rods were hammered deep into the hole base to provide extra strengthening. The relatively small (40 cm; 16 inch) plinth height ensures that the telescope is always easy to use from a seated position, wherever the object is in the sky. In addition, checking the corrector plate for dew during an observing session, even with the dew cap in place, is achieved without using a tall ladder, even if the telescope is pointed at the zenith.

Figure 3.1. Excavation of the hole for the plinth.

Figure 3.2. The plinth in position.

A key consideration in the plinth design was the interface between the plinth and the Meade Superwedge which tilts the Schmidt–Cassegrain fork so the telescope is equatorially mounted. The base of the Meade Superwedge is designed to interface with the Meade field tripod, so, not surprisingly, the top of the plinth needs to resemble the tripod head. There are five interface points which need special consideration, namely: the central hole in the Superwedge base through which a large threaded rod passes; three button head screws which pass through the Superwedge to the tripod/plinth head; and the interface between the Superwedge azimuth thrust bar pin and the tripod/plinth tangent arm. By measuring the crucial dimensions for these interface points on the Super-wedge and passing the resulting engineering drawing to a local engineering firm, the baseplate shown in Figure 3.3, atop the plinth, was produced. Metal fins under the baseplate enabled it to be pushed firmly into the wet, setting concrete, inside the plinth, during final assembly. Figure 3.4 shows the Superwedge mated to the baseplate. Shortly after this phase, the required 13 metre channel was dug between house wall and plinth, enabling telescope control data, image data and mains power to travel between house and telescope plinth.

Figure 3.3. The baseplate atop the plinth.

Shed Design and Assembly

Experience gained from my first two run-off sheds had taught me that, for minimum hassle, a run-off shed needs to be a very rigid structure, despite the fact that it must obviously have no floor (and a missing wall when the door is open!). In addition, the rails must be perfectly level and parallel and the shed castors must be precisely spaced and well greased. I cannot over-emphasize how much you will regret it if these points are not fully addressed. Trying to push a flexing, jamming, rusty-castor run-off shed back into place at 4 am when you have a bad back is no fun!

The compact tube length of the Schmidt–Cassegrain design enables any run-off shed to be a fraction of the size it would need to be for a Newtonian of similar aperture. Indeed, for a 30 cm LX200, my shed dimensions are a mere 1.2 × 1.2 × 2 m (3 foot 11 inches × 3 foot 11 inches × 6 foot 6 inches) in height, sloping to 1.8 m (5 foot 10 inches) at the back. This easily accommodates the LX200 tube length, even with an ST7 CCD and telecompressor attached.

Thus, the shed floor plan is essentially square and much more rigid than a longer structure. An additional advantage of a shed which is only 1.2 m wide is that the

Figure 3.4. Meade Superwedge attached to the plinth.

shed can be rolled back using the left hand on the left of the door frame and the right hand on the right door frame. With my large run-off sheds being 2 metres or more in width, all the force must be applied on one side, and with a semirigid shed, the whole structure twists and tends to jam when it is on the move.

As with my previous sheds, I had the Meade shed component parts made to order by a professional carpenter and shed manufacturer. Detailed plans were supplied and the shed panels (four walls and a roof) were delivered to the site prior to final assembly. Despite the extra cost, I prefer to employ a professional in this way as you end up with a quality building which will last a lifetime and will be rigid and resistant to extremes of weather. Home-made run-off sheds seem to often feature sagging and leaking roofs after their first encounter with the British winter!

In both my previous run-off sheds I had employed four turnbuckles (see Figure 3.5), two on the east side and

Figure 3.5. Turnbuckle arrangement on the south-west corner of the shed.

two on the west side, to lock the stowed shed to the wooden rail supports when the telescope was not in use. This worked fine at holding the shed down (even in storm-force gales) but there was considerable hassle in securing and releasing the turnbuckles furthest from the door; one had to crawl, in the dark, down the length of the shed, past spiders the size of small dogs, to get to the furthest turnbuckles. I tried a different approach with

Figure 3.6. Rod and hoop stowage system under the north shed wall.

the LX200 shed. The underside of the north wall of the shed features two metal hoops which hang beneath the shed and mate with short metal rods when the shed is pushed back into the stowed position (Figure 3.6). Using this system, turnbuckles are only needed on the south end of the shed, within easy reach once the door is open.

Getting the Rails and Wheels to Work in Harmony

As already mentioned, levelling the rails and ensuring they are parallel is crucial. The approach we adopted, after learning with two previous sheds, is as follows:

1. Dig two shallow channels along the proposed line of each rail and fill each channel with a single row of house bricks, flat side up. Adjust the height using soil or sand until the right and left house brick lengths are not sloping and the heights of the two rows are identical (use a spirit level on a beam between the channels to achieve this). The length of each channel needs to allow the shed to roll back so it is well clear of the telescope. Four metres is a sensible length for the rails of a small observatory. The shed width and how/where the castors will fit under the shed is crucial to the inter-rail distance; this needs detailed consideration before the lawn is excavated!

2. Treat two lengths of timber, in this case 10 × 5 cm (4 × 2 inches) in cross-section, with wood preservative and lay one on top of each row of bricks. Ensure that the inter-rail distance is roughly as required for the shed width.

3. With a hammer or sledgehammer knock metal angle iron lengths, about 25 cm (10 inches) long, into the ground at each end of each timber rail such that the right-angle folds around each corner of the timber. Screw the top of each angle iron strut to the end of each timber beam (predrilling the angle iron holes prior to burial helps here!). Attaching a third angle iron post in the middle of each timber length makes the arrangement far more rigid. You now have two timber rail supports which are the same height along their length and are not sloping. The final stage is to rest the metal rails on the timber supports ready for

testing with the shed in position. Lightly pinning the rails with nails through predrilled holes is of great help. The rails themselves are 4 metre (13 feet) long lengths of inverted "T" profile angle-iron, i.e. "V" groove pulley wheels simply run on the edge of the upturned "T". Such lengths can usually be found at scrap-metal dealers.

4. At this stage the shed needs to be assembled. The technique here is to attach large free-running pulley wheel blocks to the base of each shed wall first and then join the shed sides together with the rear wall and, finally, attach the door frame and roof. It helps if the shed sides can be assembled in situ, i.e. with the pulley wheels on the lightly pinned rails. For large sheds, three pulley wheels per side is better than two.

5. Once the shed is up on the rails, trials can be carried out to assess the freedom of movement of the shed and, after adjustment, the rails can be firmly screwed into position.

Rails, shed and wheels should now be working in harmony.

I like to add a plastic skirt to the base of the shed at this point to minimize ingress of dirt and tarantulas into the observatory building.

The penultimate task is to attach the shed door plus a hook and eye to keep the shed door in position when the telescope is in use and to stop the door being a nuisance in windy weather.

Finally, the shed was painted a pleasing shade of weatherproof green, so it blends in nicely with the garden (Figure 3.7).

Remote Control of the LX200; A Short Guide to Wiring Hassles

Despite the attractions of the LX200 for remote observing and the software available for telescope control, I could find no suppliers of PC serial port to LX200 interface cables! Even the LX200 manual has the minimum amount of useful data on this and related issues! Essentially, pins 2, 3 and 5 on the nine-pin PC RS 232 port need to go to pins 5, 3 and 4 respectively, on the LX200 port marked RS 232. You need to make up your own cables, interpret

Figure 3.7. The finished shed in position.

the diagrams in the Meade manual, and acquire the necessary nine-pin and phone jack connectors!

The PC's parallel port connects to the CCD camera. If you are using, say, an autoguiding ST7 CCD, you need to plug the ST7 guide lead into the LX200 CCD socket on the Meade control panel. If you want to control an LX200 electric focuser through the RS 232 lead, using, for example, *The Sky* software you need to plug the electric focuser cord into the focus socket on the Meade control panel. If you are planning on doing a lot of remote observing you will need focuser control as the LX200 focus shifts when the telescope slews across the sky. Also, don't forget to plug the LX200 declination motor cable into the LX200 declination socket.

The Observatory in Use

The observatory has been in position now for three years and is a joy to use. The ease with which this small shed rolls back and the speed with which the telescope can swing into action has transformed my observing from a tedious chore to a pleasure. It really is amazing what a difference a user-friendly observatory makes. Since the initial design, I have added a few extra features, namely:

1. A small work-surface on the north side of the plinth, for resting charts and observing data.

2. A small shelf for similar items inside the shed.

3. A red light inside the shed, above the doorway, to give a pleasing dim red glow, via which charts can be read.

4. Rubber mats around the paved area surrounding the plinth, so kneeling is less painful.

5. Extra zero-power and illuminated right-angle finders for initially centring the telescope on a known star.

6. A Black & Decker Snakelight, which wraps round the observer's neck, providing hands-free use of a torch when observing – a superb gadget!

Using the telescope remotely works well, but, personally speaking, I prefer to be outside with the telescope. Despite being primarily a CCD observer, I feel I am "missing the point" if I am an astronomer and am not outside in the cold, clear, night air, able to see bright meteors, use binoculars, let some original photons hit my retina and watch the progress of the inevitable cloud. Also, even using a dew cap and dew-zapper heated band

Figure 3.8. The author, ready for observing.

Figure 3.9. The shed fully rolled back to the north.

is not always enough in the damp British climate. Being on hand with a powerful hair dryer and a torch to inspect the corrector plate is a good plan. It can be intensely frustrating to be indoors and not know whether your CCD images are deteriorating due to cloud, dew or some other factor (e.g., the telescope trying to see through a tree!). I also feel distinctly like a couch potato if I am sitting in front of a monitor all night (it's something I do all day at work too).

Finally, I must acknowledge my father's invaluable help in all my astronomical endeavours, especially observatory building. Without his professional DIY approach my sheds would probably never have been built and would certainly have looked distinctly "Heath-Robinson" in appearance!

Chapter 4
Darklight Observatory, Eddyville, Kentucky, USA

Chris Anderson

My adoration of the skies began at an early age. I can recall, being ten years old at the time, dragging my Edmund Scientific 6-inch reflector from the den and out to the driveway. Sometimes hours would pass before I would retire, wind-chapped and frozen, but with a mind full of stars. Some years later, after my aging "6" was replaced with an orange Celestron 8, I dabbled in astrophotography. This was extremely limited by the slow emulsions of the day, and my neighbors' incessant use of porch and sodium vapor lights. I often dreamt of my own quiet, dark place, upon which I could build my own observatory. My family even talked of buying a parcel of land far from the Louisville city lights that surrounded me, but this never came to fruition. My interest in astronomy waned somewhat in later years, but my dream of an observatory never faded.

After becoming employed with the Kentucky State Police, I soon found myself living and working in remote western Kentucky, far from the glow of city lights that had so plagued the astronomical endeavors of my childhood. I had purchased a home upon several acres of rural farmland, and realized that the time had come to build my astronomical observatory. My interest in astronomy had been rekindled by recent advances in amateur CCD imaging. I saw this as a definite revolution over film, with its generally shorter exposure times and easy digital manipulation with computers.

I had studied many telescope systems and CCD cameras, as well as mounting platforms, and finally decided on a Celestron Fastar 8 and SBIG ST237 CCD camera. It would be supported by an Astro-Physics 900 GTO goto mount. As there was a nine-month wait for the Astro-Physics mount, I had ample time to design and build the observatory. My goal was to have the structure complete before the arrival of the mount.

Construction officially began in June of 1998. Having no prior experience with carpentry, I enlisted the help of a close friend to assist in the building. I had a basic idea in my head of what I wanted in an observatory, and only a few rough sketches to guide me. My prerequisites were a roll-off roof, ample space for a desk and chair, and walls high enough to stop wind, but low enough for the scope to have an unobstructed view of the sky. Rather than have a door one would have to stoop to enter, I decided to go with a full-height exterior doorway. This mandated the walls being fixed height, so all other measurements (concrete pier height, etc.) were based on this height. I could have cut a regular exterior door down to a smaller size, but this would have been much more difficult than buying a pre-hung door "off the shelf". Plus, I wanted the extra security of a ready-made door.

Having just finished adding a deck to my home, I decided to use the same construction technique on the observatory. Rather than pour a large concrete pad upon which the building would sit, I would build an 8 × 10 foot (2.4 × 3 m) deck frame. This would be much easier than messing about with concrete. It would also, I hoped, be less of a thermal problem.

My immediate problem was to determine how high the central pier support for the telescope would have to be in order to see over the walls. Since I had taken the easy way out by using a standard door, I had to be sure the scope could see over the nearly 7-foot-high (2 m) walls. Stability in such a tall pier was a concern, so I decided to use a LeSuer Astropier to support the mount and scope. I purchased a rather short 36-inch model. The difference needed to clear the wall height would come from a concrete column. A section of 18-inch wide Goedecke tube left over from a Dobsonian project would do nicely for a concrete form. A square form made from plywood could also have been used.

The first stage of construction was to dig a hole for the pier. I wanted a pier so solid that my only fear would be earthquake. To dig such a hole, I enlisted the aid of a friend who had a large auger truck. He was able to drill a 5-foot-deep (1.5 m) hole with little or no effort (Figure 4.1)! LeSeur suggested a 3-foot-deep (1 m) hole

Figure 4.1. A truck fitted with an auger bit is used to dig the hole for the concrete pier.

to set their threaded rebar pier attachments into. My friend, who sets bridge piers for a living, suggested five might be better, to leave room for steel reinforcing rods (which he kindly donated to the construction).

After the shaft was sunk, I constructed a simple form around the opening. Based on my earlier calculations, the height of the concrete column would be enough to raise the total height of the scope to just above wall level. This worked out to be about 3 feet (1 m). I used a 4-foot (1.2 m) section of Goedecke concrete form tube, placing about a foot of it below ground level. The whole tube was very securely held in place with 2 × 4 inch (50 × 100 mm) beams and old landscaping timbers. One must never underestimate the potential energy of wet concrete. I had to make sure there was no possibility of collapse of the concrete form as concrete was being poured into it. Since the inside diameter of the hole was slightly larger than the form, the base of the tube was fitted with a plywood skirt. This would serve as a dam to keep concrete from welling up around the tube (Figure 4.2). The tube was leveled before being staked in place. In the image you can see the LeSeur threaded rebar that will be placed into the top of the still-wet concrete column.

I calculated I'd need approximately 1.5 yards (metres) of concrete to sufficiently fill the form. This was much more than I cared to mix from bags, so I opted to have a cement truck come in (Figure 4.3). Plus, having it trucked in alleviated any possibility of the concrete setting in layers while I mixed bags. Steel reinforcing rods were placed down inside the tube, and

Figure 4.2. Sonotube concrete form and support timbers ready for pouring. The threaded rebar suite for the LeSeur Astropier can be seen to the right.

Figure 4.3. Trucked-in concrete being poured in to the Sonotube.

I made sure they would not interfere with the setting of the LeSeur rods which would be pushed into the concrete. It's easy to panic when trying to get these rods down into the concrete. At first, I didn't think they would go in due to the gravel content. But with a little work, they seated flush with the concrete surface. It is worth noting that one of the three LeSeur rods needs to be set roughly due north. After the concrete had sufficient time to set, the supports were removed and the form cut away. So that I could use the pier with my existing telescope, I made a temporary adapter plate that would allow the fork-mounted OTA to be placed on the pier (see Figure 4.4).

The next step in construction would be to mark and dig holes for the decking supports. Using the same technique I used in the construction of my house deck, I drilled several holes with an engine-powered post-hole digger. The manual variety could be used, but my choice of locations was atop an old road bed. The foot-thick layer of gravel negated the use of a hand tool (Figure 4.5)!

Figure 4.4. Finished concrete pier and fork assembly attached with adapter plate for testing.

After digging the post holes, 4 × 4 inch
(100 × 100 mm) posts were set in concrete and
accurately aligned using batterboards and string. It
took about a half a bag of concrete to set each post. This
can easily be mixed by hand in a wheel-barrow.
Figuring where the posts should be set was a nightmare.
The pier is offset from the middle of the floor (so more
room would be available to accommodate a desk), and
its mere presence made it impossible to accurately
measure diagonally to check squareness. The old
carpenters 3-4-5 rule helped, as did a bit of luck
(see Figure 4.6)!

Next a simple beam and joist structure was built
around the posts (Figure 4.7). The first outside beam
was carefully leveled at one end, then each subsequent
beam lined up and leveled at the opposite end. This was
done all the way around the outer posts so that when I
came back around to where I had started, it was off by
6 inches (15 cm)! I guess my level was not accurate
enough. Nevertheless, after a bit of eyeballing and
tinkering and leveling, it all worked out. The beams
were in place and could be bolted to the posts with
4-inch (100 mm) lag bolts. Then the excess post height
was cut away. The plywood flooring was added next,
and some wiring placed underneath. Since the pier is
offset from the center of the observatory floor, I needed

Figure 4.5. Digging
the holes for the
buildings corner posts.

Figure 4.6. Aligned with string, the posts are set in concrete. Later they will be cut to the proper height.

to cut a precisely centered hole in one piece of flooring to slip over the column. A bit of very careful measuring made this feat possible (see Figure 4.8).

Once the floor was complete, stud walls could be erected. As previously mentioned, the wall height was determined largely by the height of the door. So, studs were cut accordingly. Three of the walls were constructed under my carport, carried to and placed upon the floor (Figure 4.9). The fourth "door wall" was constructed after the door was placed. Do not try to make the door stud wall first. There is too much chance the pre-hung door won't fit properly! During this phase

Figure 4.7. Beams and joists are attached to the posts, in much the same way as a deck is built.

Figure 4.8. Completed decking ready for the wall construction.

of construction I was faced with several dilemmas. I found out the building thus far was not perfectly square, and that hanging a door is not a one-person job! It helps to have friends around for this. After the door was installed, wall covering could begin. I opted for T-111 exterior paneling. It features a type of tongue and groove interlocking edge, and generally requires no bonding or caulking. It was simply cut to fit and nailed in place (Figure 4.10).

The roof section was then constructed. First, simple sloped roof trusses were constructed on the ground. I

Figure 4.9. Construction of the standard stud walls was completed under the carport then carried into place.

Figure 4.10. Lastly, the door was placed and studs erected around it. Then exterior paneling was added.

knew early on the roof would be heavy; too heavy, in fact, to lift into place without a crane. So, the roof was built, *sans* rolling tracking, directly atop the building walls. After the trusses were put in place, all supporting and connecting rafters were added, then additional T-111 paneling was cut to fit. The paneling was allowed to extend down past the level of the walls in order to block wind and rain from blowing over the walls

Figure 4.11. The roof rafters were constructed in sections and then placed atop the building.

(Figure 4.11). Later, several sheets of corrugated metal roofing were screwed securely in place.

With the roof construction complete, it was time to add the rolling tracks. One of my great fears was the roof being lifted up and blown off during high winds. I live in the outskirts of "Tornado Alley" and I knew nothing would protect the structure from even a small twister, but winds were a concern none the less. Originally, I had planned on using a simple "V" track and rollers. This would necessitate the roof being held down by some locking mechanism when not in use. Instead, I discovered Cannonball track at the local hardware store. Originally designed as a hanging roller track for garage doors, I found it could be mounted upside down, atop the walls. The rollers (see Figure 4.12) would then be mounted to the base of the roof structure. The captive design of the Cannonball track would lend some measure of safety in keeping the roof securely in place (see Figure 4.13). To install the tracks, one end of the completed roof was literally jacked up, the tracking installed, then the same was done to the other side. Four by four track beds were attached to the back side of the building, and additional

Figure 4.12. Cannon–ball track roof rollers are attached to the base of the roof structure.

Figure 4.13. Roof roller and Cannonball track.

tracking laid down, so the roof would have somewhere to roll to when in use.

Originally, I had hoped the roof could simply be rolled off by hand. The sturdy construction, weight of materials, and the high center of gravity made this

Figure 4.14. Hand-powered winch used to roll the roof off. A second winch located on the other side of the building is used to close the roof.

impossible without a ladder. Even then, it was hard to get any leverage. So, I decided to install two hand-winches, one fore and aft (see Figure 4.14). The "open" winch uses a pulley and steel cable attached to the outside of the rear wall. It is used to easily open the roof, in effect pulling it rearward. The winch itself is installed inside the building, with the cable exiting through a hole in the floor. To close the roof, a cable is manually attached to the center roof rafter and cranked shut. When closed, both winches can be tightened down to increase security and weather-proofing.

After completing the construction, the LeSeur Astro-pier was bolted securely to the top of the concrete pier (see Figure 4.15). The interior of the observatory was carpeted to help ward off cold and the walls were painted flat black. Wiring and data cables, and power cabling, were installed, along with a work bench/desk. Finally, the Astro-Physics mount and Celestron Fastar OTA were permanently attached. When parked (with the declination axis and the right ascension axis level) the roof rafters neatly miss the contraption when the

Figure 4.15. The mounted LeSeur Astropier.

roof is removed. I did pad the wall height measurements a few inches to accommodate a larger OTA in the distant future (see Figure 4.16).

Darklight Observatory has been in use now for several years (Figure 4.17). It has proved to offer more than adequate protection for me and my equipment during all types of weather. I have installed a dehumidifier to take some of the moisture out of the building. One problem with the design is the slight gap between the roof overhang and the top of the walls. This does allows free air circulation (good for temperature regulation), but also allows bugs to get in. Every winter several thousand ladybugs find their way in to hibernate. They don't bother me so I let them be. Future plans involve the creation of a network to control the observatory computer from in the house. All in all, for the total materials investment of US$2700, I have been very pleased with the outcome of my own astronomical observatory.

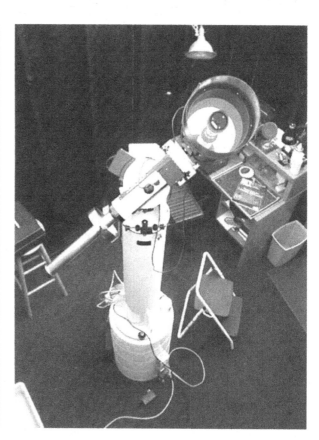

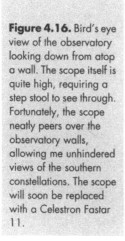

Figure 4.16. Bird's eye view of the observatory looking down from atop a wall. The scope itself is quite high, requiring a step stool to see through. Fortunately, the scope neatly peers over the observatory walls, allowing me unhindered views of the southern constellations. The scope will soon be replaced with a Celestron Fastar 11.

Additional images and text can be found on the World Wide Web: http://www.darklightimagery.net.

Figure 4.17. The completed Darklight Observatory.

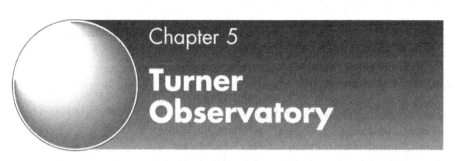

Chapter 5

Turner Observatory

Bob Turner

It all started with road works and a diversion around Brighton station. Sitting in a slow-moving line of cars around the back streets of Brighton while the rain intermittently pattered on the window is not an edifying experience. To alleviate the boredom my attention was taken by the woman in the car in front being more interested in controlling various fractious offspring than keeping up with the line of traffic.

The set of temporary lights went red and once again we were stationary having moved all of 10 feet. Looking out the window to my right was the yard of a metal fabrication business and amidst the stacks of iron and steel pipes and tubes, leaning against the wall, was something that caught my eye, huge half-hoops of angle iron. It did not immediately register, but there was my observatory. Pulling out of the line of traffic, I drove into the yard the only way I could through the "exit only" sign and parked close to the target of my attention.

There were six half-circles of 3 × 3 inch (75 ×75 mm) angle iron, some 14 feet (4.2 m) in diameter. The metal was a bit rusty but looked true in shape, as all the hoops matched each other. It did not take me long to find the manager and enquire if these were manufactured items and their price. The reply was no, they did not make them, but the metal was left over from a fabricated archway for a hotel and was scrap waiting to be sent the following week to the scrap dealer.

"Were they for sale?" "Yes."

Now the $1000 question. "How much?" "You can have the lot for £25 cash."

"Can you deliver them?" "No sorry we don't have any jobs in West Sussex."

Now here was my first big problem. I live in Worthing about 20 miles away. How was I to shift six half-hoops of steel 14 foot long and standing 7 foot high? Ignoring the problem in my excitement I promised to collect the material on Saturday morning, paid my money and obtained a receipt.

What to do now? If only I had known what I was letting myself in for. Brighton Van Hire was an obvious solution, but a visit there with a tape measure soon convinced me there was nothing I could drive that could carry these bits of metal without them sticking out of the back of the vehicle. I then thought of using a trailer with the hoops mounted upside down, but where do you find a trailer 15 feet long?

Two days later I was nowhere nearer the solution. I had the basis of my observatory but how to physically move it seemed insurmountable. It was then that fate struck a second time.

One of my astronomical associates works for the Ministry of Defence and on the Thursday evening we were at a meeting together. Explaining my problem he said that he was moving a load from Hastings to Salisbury the next day in a vehicle big enough for the job and would be glad to help. Could I be at the collection point on Friday about 11 o'clock to help him load up and then follow him back to my house and unload?

So there I was at 11 o'clock, waiting on the pavement for Steve to arrive. Five past, ten past. I was just beginning to worry, when round the corner came the biggest camouflaged low loader you have ever seen in your life, with a Churchill tank on the back. Loading was accomplished by tying the hoops to the side of the tank, but before we had finished we had collected a crowd of about 20–25 spectators. Then followed a sedate drive along the A27 behind my observatory, with me looking directly up into the barrel of a gun.

Stowing the load at the side of my house was no problem but it did attract considerable neighbourly interest. How many times have you ever had a tank parked outside your house?

So now I could start. First the base. I wanted the telescope mount to be vibration-free, so that walking on the observatory floor would not transmit shake to the telescope. To accomplish this, a hole 20 inches (50 cm) square and 3 feet deep (1 m) was opened up on the centre of the site and about a foot of wet concrete was poured into the hole. The telescope base – two bits of 8-inch (20 cm) girder 2 ft 6 inches (80 cms) long welded together with a half-inch (12.5 mm) steel plate welded on top – was lowered in until it stood 3 inches (75 mm) proud of the calculated observatory floor. The balance of the hole was then filled with concrete, the top 6 inches (15 cm) being shuttered with timber to form an 18-inch (45 cm) square, the top level with the observatory floor.

When the concrete set, the shuttering was removed and the top of the foundation was wrapped with half-inch thick heavy-duty rubber sheet which would separate the observatory floor from the mount. It was now a matter of levelling the hardcore, adding the outer shuttering to form an 18-foot (5.5 m) octagonal floor and ordering the concrete. A few days after pouring I was standing on the base of my new observatory, which had begun to look considerably bigger than my kitchen.

The rotation of the observatory was to be supplied by mounting the angle iron on 6-inch (15 cm) diameter heavy-duty plastic wheels, the bases of which were cemented into a three-brick high 14-foot (4.2 m) diameter wall built directly on to the concrete floor. Twelve wheels in all were used, spaced evenly around the circumference, positioned so they would sit within $\frac{1}{8}$ inch (3 mm) inside the lip of the angle iron loops. We had tested the loops for accuracy by laying them face-to-face on the driveway and measuring the diameter inside face to inside face at several places around the circle.

We were now ready to weld the first steel to make the dome by joining two of the half-loops to form a complete circle. The metal was cleaned, painted and the joint faces prepared. The finished circle was welded on both sides and the inner face ground to give a smooth transition from one half-loop to the next for the wheels.

Picking up a piece of metal with a radius of just over seven feet is not that difficult but moving it is something else. Standing in the middle, arms stretched, and getting both ends off the ground is difficult, but turning over a 14-foot circle is quite traumatic.

Soon the first circle was completed and four of us lifted it into place on the rollers and for the first time rotated the main runner, I must admit far more accurately than was ever expected. The first of the up-and-over steel could now be erected and the two sides of the dome slot were put in 3 feet (1 m) apart with sections of straight angle iron welded at the bottom ends

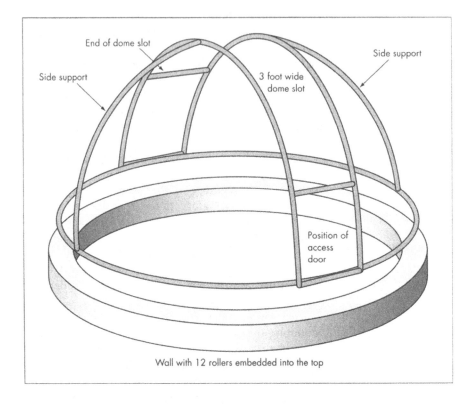

End of dome slot

Side support

Side support

3 foot wide dome slot

Position of access door

Wall with 12 rollers embedded into the top

Figure 5.1. The steel structure.

and part way up the structure. These cross-pieces were for strength while assembling and to take the top of the door and the back end of the dome slit. A fifth loop was used, cut into two smaller lengths as side supports.

The next task was to find a material to fill in the dome panels to carry an outer sheath of fibre-glass. Initially I wanted to weld in thinner struts to give some support to the large segments of dome and strips of 1 inch × $\frac{1}{8}$ inch (25 × 3 mm) cold rolled steel were surprisingly inexpensive. The last half-hoop was used as a template and the cold rolled steel was bent by hand into the curve desired and then welded to the structure. Cross-braces of the same material were put in to form a geodetic structure and eventually there was a cross-brace every 18 inches (45 cm).

The dome looked like a great big skeleton and to my eye totally massive. An open evening saw 15 people inside the structure at the same time without being squashed and the top of the dome was now well above outstretched hand height. I was beginning to get nervous. It's fine having things on paper, but to

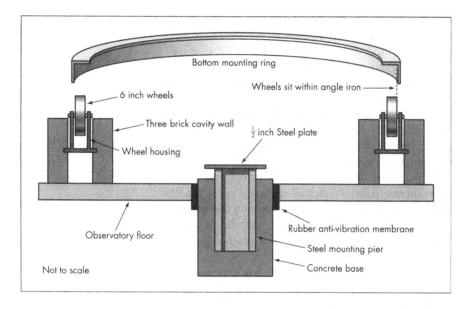

Bottom mounting ring

Wheels sit within angle iron

6 inch wheels

Three brick cavity wall

$\frac{1}{2}$ inch Steel plate

Wheel housing

Observatory floor

Rubber anti-vibration membrane

Steel mounting pier

Not to scale

Concrete base

Figure 5.2. Floor detail.

physically see them represented in steel makes you stand back and wonder just what you have taken on.

The membrane material between the cross-struts needs to be rigid enough to support an overlay of fibre-glass but flexible enough to take up the spherical shape of the dome. So what to use? I experimented with lots of material but the end-result needed to be a material that was cheap and readily available, as there are many square feet to cover in a 14-foot half-sphere.

Plywood was too rigid, hardboard split when bending, polystyrene was too flimsy, cardboard went soggy before fibre-glass set; the choices were endless but the solution unreachable. At this point the observatory stalled for four to five weeks. I was in our local garden centre and saw that among the wood they were selling off cheap there was several sheets of formica laminate marked down to £1 each. Swiftly snapping up this bargain the sheets were trimmed to interlocking panel sizes by scoring and snapping along the score line. Then started the long process of carefully drilling $\frac{1}{8}$-inch (3 mm) holes and pop-riveting the panels to the cross-supports. The drilling of laminate is difficult but can be achieved with care.

About three weeks' work went into covering all but the dome slot with thin laminate sheets, but soon it was completed and the dome really looked like an observatory. From a local manufacturer of channel buoys and boat moorings I purchased resin and hardener and from the liquidator of a local fibre-glass company several rolls of 12-inch (30 cm) wide fibre-glass. We were now ready to start the finishing touches.

The gods had really been on my side up to now, but this was the autumn of 1987. I went to bed with a fully clad observatory and awoke to find the structure was back to being a denuded skeleton again. The great storm of 1987 got the observatory, all our garden fences, many roof tiles, our greenhouse and a young tree we never saw again. It completely disappeared.

All the trees were down in the street and it took several hours before we cleared sufficient space to allow vehicles to gain access. On the observatory there was not a shred of Formica laminate left – just the pop-rivets, although we did share the bits with neighbours up to seven or eight gardens away.

The financial loss was minimal but the work had taken forever. So back to the drawing board. The garden centre had no more of the same material and the thought of drilling all those holes that had taken so much care was daunting. Then a stroke of genius. What about offset litho plates? They are thin, robust, easily drilled and bent into shape. Why didn't we think of them earlier!

A visit to a local printer and I was the proud possessor of enough plates to make two observatories all at the princely cost of nothing, an amount I try to specialise in.

With three helpers the observatory was clad in one weekend and the next weekend saw the fibre-glass covering put on. The use of such a malleable material allowed a skirt to be put round the bottom to act as a drip ledge, and with a heavy-duty plastic sheet spread across the dome slot the whole structure was water-proof, dry and totally magnificent.

The last half-circle of steel was again pressed into service as a template for the dome slot cover which was manufactured from 1½-inch (38 mm) steel angle, cross-braced and welded and then given a covering of aluminium offset litho plates and a final covering of fibre-glass. The runners for the slot cover which went off to one side were made from a length of 3 inch (75 mm) square section aluminium tube cut in half on a band saw, and the slot cover itself ran in these tracks on shopping trolley wheels kindly donated by Tesco.

A porch of 1½-inch (38 mm) welded steel angle was added to the structure and a double-glazed window salvaged from the scrap-yard made a 3-foot (1 m) square door. There only remained a coat of paint inside and out and the structure was finished. The observatory housed a

Figure 5.3. The completed observatory.

10-inch Newtonian and a 12-inch classical Cassegrain mounted parallel on a steel-fabricated fork mount.

All this was some 15 years ago. The observatory has had considerable use and still remains as sound as it ever was (Figure 5.3). Alas, however, its owner has not suffered so well and now finds the effort of such a large construction a bit daunting.

Chapter 6
A Simple Rotating Observatory in Nottingham, England

Alan W. Heath

Due to a change of address, it was necessary to build a new observatory in 1997. The original was described in *Small Astronomical Observatories* (1996) and the same basic design was followed. Previously the sides of the building were made from sheet asbestos, but 12 mm ($\frac{1}{2}$-inch) external plywood sheets are a more acceptable alternative, these being mounted on a 50 × 50 mm (2 × 2 inch) timber frame. The sides and the roof are held together with bolts, thus allowing the observatory to be dismantled, should the need ever arise.

A permanently mounted telescope needs protection from the weather, and living in an urban area calls for screening from local lights. A proper observatory also offers some degree of comfort in the wind and cold of winter.

Domes always present construction problems, so I decided on a design rather like a simple shed but with a hinged roof section to permit access to the sky. A hinged extension to the shutter on the same wall as the shutter permits observation at lower levels if required. Even lower objects can be observed through the open door! The fact that the walls are flat rather than curved is an added convenience that allows for the permanent fixing of charts, photographs, maps, etc.

Rotation of the Observatory

The entire building rotates. It is mounted on an angle iron ring 2.45 m (8 feet) in diameter and this was salvaged from the previous observatory, together with the eight wheels upon which the building rotates. The pulley wheels are 75 mm (3 inches) in diameter and have 12 mm ($\frac{1}{2}$ inch) bolts as axles. The ring sits on the wheels, which face upwards (Figure 6.1).

The angle iron ring was the only part not made at home. It was necessary to make enquiries at several local engineering firms to find one, which had facilities to roll the angle iron ring and weld the joint. It is made from 50 × 50 mm (2 × 2 inch) angle iron, flange outwards. It is painted with red metal primer and has a top coat of bitumen paint.

The base of the observatory is a fixture. It is a square frame made from 230 × 50 mm (9 × 2 inches) timber, with angle pieces made from 75 × 50 mm (3 × 2 inch) timber placed across each corner, providing eight points that are equal distance from the centre. The wheels are placed at these points and the angle iron ring lowered on to them. The wheels have to be carefully adjusted before finally being secured to ensure the ring turns easily.

The wheels must be level and the bolt, which is the axle, needs to be half as long again as the wheel is wide in order to allow the wheel to "float" and so

Figure 6.1. Pulley wheel upon which the circular angle iron track rotates.

compensate for any minor errors in the circularity of the ring itself. The observatory building is secured by screws through holes in the ring.

The wheels and axles are supported by short lengths of angle iron on each side, the axles passing through holes in these. The whole wheel assembly is further mounted on a piece of 150 × 150 mm (6 × 6 inch) sheet steel and then fixed in place on the wooden base. This makes easy any adjustments before finally placing the ring in position.

Lubrication is achieved with a mixture of graphite and grease which is efficient and quiet – an important point when using the observatory in the small hours of the morning! No signs of wear and tear have been noticed in either wheels or axles.

The Base

The observatory is on a concrete base which is about 3 m (10 feet) square, thus providing a slight overlap. The base is approximately 150 mm (6 inches) deep. A cubic metre (35 cubic feet) of concrete was used for this, together with some aggregate. In practice it has been found that the observatory can be moved easily by hand, but it does not move of its own accord even in a strong wind.

Actually there is no need to have a base unit at all as the wheels could have been fitted directly to the concrete, but there is the advantage of having some clearance of around 450 mm (18 inches), so lifting the observatory clear of any snow on the ground without much trouble. Snow on the roof can usually be removed just by opening the roof shutter (Figure 6.2).

Mains electricity enters the building via a conduit through the base connecting first to a switch box. Electrical safety regulations must be met when working outdoors with mains electricity. A circuit breaker is a valuable safety feature.

All cables for lighting and so on run along the inner walls from a junction box near the centre of the roof. From this the cable is fitted with a male–female connector to permit disconnection and the periodic removal of "twists" caused by the rotation of the building more often in one direction than in the other.

Additional fittings include a drop-leaf table-top in one corner for charts, a battery-operated clock showing

Figure 6.2.
Observatory showing the building in relation to the fixed base. The main roof shutter and the small lower shutter are shown in the open position.

universal time and a small cupboard for various accessories such as eyepieces, filters and other items.

Using the Observatory

For security reasons, a battery-operated alarm is fitted. The observatory has a permanently mounted 250 mm Newtonian reflector (Figure 6.3). It is used mainly for lunar and planetary observations, which I contribute to the various sections of the British Astronomical Association as well as to overseas organizations. A 75 mm Broadhurst Clarkson refractor with sun diagonal is also used for solar observation.

The design has proved to be very efficient, the first one operating trouble-free for some 36 years. The only maintenance needed is to lubricate the wheels annually and to apply wood preservative to the woodwork from time to time. No replacements have been necessary.

I would not change this design – it can be scaled up or down to suit the size of telescope. The one described has the advantage of using standard-sized plywood sheets. Two 2.4 × 1.2 m (8 × 4 feet) sheets cover one side so there is minimum wastage. A design of different dimensions may result in some wastage.

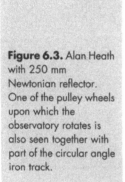

Figure 6.3. Alan Heath with 250 mm Newtonian reflector. One of the pulley wheels upon which the observatory rotates is also seen together with part of the circular angle iron track.

Anyone who is remotely handy can construct a building like the one described, which has the advantage of looking like an ordinary garden shed, and is therefore "neighbour-friendly", while retaining the benefits of more sophisticated designs.

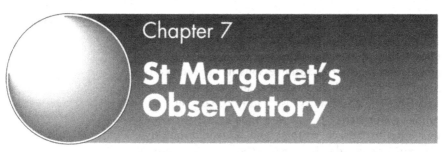
St Margaret's Observatory

Paul Andrew

After three decades of owning a variety of telescopes from 6-inch to 10-inch Newtonians and Cassegrains, I finally decided to go for a large instrument, which would fully satisfy my thirst for a wide range of deep-sky objects. A Meade Starfinder 16-inch, f/4.5 equatorially mounted Newtonian reflector was duly acquired. After struggling to assemble the mount and tube before each observing session it soon became very obvious that the expenditure would not stop with the instrument, but that a form of permanent housing would be a necessity and not a luxury (Figure 7.1).

After much deliberation, and research on the Internet, a run-off roof observatory was finally decided

Figure 7.1. The completed observatory with the roof fully rolled back.

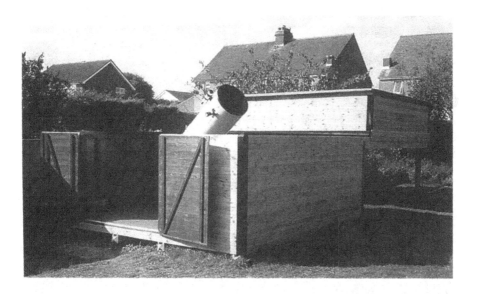

upon. This design provides better cooling to the ambient air, and for security reasons its appearance is more like a normal shed rather than the classic dome structure. The observatory also had to be large enough to take a number of people at a time as the telescope was sure to attract visitors, and it would quickly become very crowded in a small shed.

The size of the observatory was therefore set at an optimum 10 × 12 feet (3 × 3.6 m), with a pent roof running on strong castors, supported on 4-foot (1.2 m) high walls. Ideally, the roof should roll off to the north, but as my 190-foot (60 m) garden runs in a north-west–south-east direction, a north-west run-off would have to do.

While the general features and dimensions of the observatory were carefully considered in advance, many of the details would be sorted out as the construction developed – in a sense there would be an "organic" evolutionary component to the design. This I hoped would give a certain amount of flexibility when overcoming problems. While this approach may not be suitable for everybody, it did prove successful in this instance.

The 10 × 12 foot (3 × 3.6 m) base was pegged out after careful consideration was given to the best location of the observatory to ensure a maximum visible sky. A further 12-foot (3.6 m) length would be

Figure 7.2. The ¾ –inch (18 mm) shuttering is cut to size and screwed to the framework. Note the 6–foot (1.8 m) concrete posts acting as the foundation for the observatory.

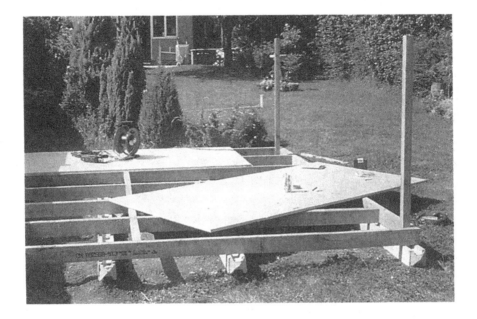

Figure 7.3. A close up of the pier embedded in its 3–foot (1 m) ball of concrete. Notice the removable hatch in the floor and the offset of the plastic collar.

required for the rails, making the true footprint of the observatory a large 24 × 10 feet (7.2 × 3 m). An old, dying hedge that ran by the side of the observatory had been up rooted and replaced by 4-foot (1.2 m) high fencing a few weeks before, and a small group of conifers had to be moved to a new location. As the garden slopes southwards it was necessary to build up certain areas of the ground while lowering other parts.

The next stage was to carefully mark out the optimum position of the pier. Once this was done a 3-foot (1 m) deep hollow ball was dug 5 feet (1.5 m) in from the doors, and equidistant from the side walls. This offset left an additional space towards the north-facing wall that could be used to house all those accessories that are needed to use the telescope to its maximum efficiency. Initially, I built a simple worktop along the wall, but this was quickly replaced with a surplus kitchen unit and top that a friend had no need for.

The initial option of a solid concrete base to the observatory was rejected so as to help reduce air currents during the summer months. Instead, eight 6-foot (1.8 m) concrete posts were laid horizontally to form the base of the foundation of the observatory.

Levelling of the posts was then carefully carried out, checked and rechecked with a spirit level, to ensure that the slope of the garden was fully neutralized.

The floor sills consisted of 4 × 2 inch (100 × 50 mm) pressure-treated timber screwed together, with each joist placed at 2-foot (60 cm) intervals. This ensured that the floor would be solid with little or no bounce when walked upon. This framework was initially assembled on the patio and then man-handled down to the final location.

Two-inch by two-inch (50 mm × 50 mm) upright corner posts were screwed into place to which the completed wall frames would be attached at a later stage. Shuttering $\frac{3}{4}$ inch (18 mm) thick was then laid down for the floor (Figure 7.2). A removable hatch was built into the floor so that access could be gained to the base of the pier should the need arise. It is vital that the pier is totally isolated from the observatory so that no vibrations are transmitted to the telescope. The floor and foundation of the observatory must therefore not touch the pier at any point.

Careful consideration was given to the height of the pier above the observatory floor. This needed to be tall enough so that the telescope tube would not hit the floor, but low enough to ensure that the eyepiece would not be placed too high for access when the telescope was near the vertical. The height also needed to allow enough clearance to close the roof by swinging the Dec axis and tube assembly into the horizontal position. A height of 19.5 inches (approximately 50 cm) was finally decided upon.

The pier itself consists of a metal pipe with a "cup" of the correct diameter to enable the equatorial head to be attached. After embedding the pipe and a thick plastic outer collar in a 3-foot (1 m) ball of concrete, the gap between the two was then filled with the remaining concrete to produce a totally stable arrangement. It was found that the collar needed to be offset to the north. This was to ensure that the motor and worm wheel housing on the polar axis would not hit the pier when the equatorial mount was adjusted to the correct latitude of 51 degrees north.

The frame of the four walls was then constructed from 2 × 2 inch (50 × 50 mm) pressure-treated timber, with 2 inches (50 mm) separation for the wall studs. Four feet (1.2 m) was deemed to be the ideal height for the walls to allow the telescope access down to the local horizon, particularly for the southern view. In addition

Figure 7.4. View (looking west) showing the wide aperture for the split door, and the slope of the roof.

the wide door, which is asymmetrically split into two sections, could also be left open to allow total freedom to this region of the sky.

Tongue and groove was chosen as the cladding in preference to shiplap. Although somewhat more expensive I felt the overall appearance of the observatory would be enhanced, as would the overall quality of the construction – it was not my intention to skimp on

Figure 7.5. View (looking due south) showing the two lengths of angle iron (painted black) screwed into place on top of the 4–foot (1.2 m) high walls and rail supports.

materials. The completed cladding was then finished with several coats of a high-quality, clear yacht varnish. Sadly, the yacht vanish did not stand the test of time and within a year the observatory looked very weather-beaten. While I am not sure why this should happen, one possibility is that some residual dampness in the "tongue and groove" seeped out over a period of time to create the weathering. The only option was to completely sand off the varnish and apply a new coating. This time I decided to use an exterior extra-durable woodstain in a semi-gloss Redwood. This has had the effect of generally darkening the observatory and to date seems to be doing the job more effectively.

Figure 7.6. The tongue-and-groove boards have been tacked into place on the four walls and the roof.

The roof runs on two specially welded 24-foot (7.2 m) lengths of 2-inch (50 mm) angle iron. A great deal of time and effort was devoted to ensure that these rails were absolutely level and totally parallel to each other. Any inaccuracy at this stage would result in future problems with rolling the roof on and off. In order to accommodate the full length of the rails, a 12 foot (3.6 m) extension was built using 6 × 2 inch (150 × 50 mm) timbers. This in turn was supported by 4 × 4 inch (100 × 100 mm) uprights held in position using "Metposts". Cross-bracing was then added to ensure overall rigidity.

The 2 × 2 inch (50 × 50 mm) pressure-treated skeletal roof section with 2 × 4 inch (50 × 100 mm) rafters was fitted with eight strong casters (four on each side) and then placed into position. Although a lot of

Figure 7.7. A close-up of one of the guide rails. Notice that the 6 × 2 inch (150 × 50 mm) support has been covered with two lengths of tongue and groove.

Figure 7.8. Detail of one of the eight castors that were simply screwed on to the bottom of the roof frame.

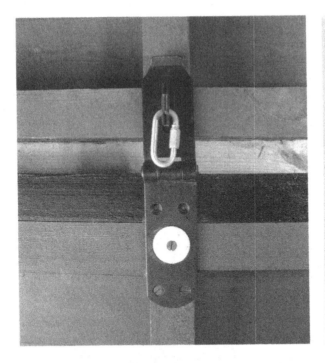

Figure 7.9. A large hasp and staple on each side secures the run-off roof to the observatory base. This system provides a very quick method of unlocking and locking the roof into position.

the construction process could be achieved with minimal assistance this phase of the operation required a person on each corner of the frame. The alignment was then checked for accuracy of construction. It was also confirmed that the drop of the pent roof (towards the west) of only $2\frac{1}{2}$ inches (6 cm) would be enough to stop any rainwater pooling on the roof.

Once satisfied that all was well, half-inch (12.5 mm) shuttering was cut and screwed into position on the roof, and tongue and groove tacked to the four sides of the structure. Additionally, an overhanging skirt of two glued tongue-and-groove boards was added to the 12-foot (3.6 m) long sides to cover and weather-proof the gap created by the height of the castors – these would also act as guides during the roll-off process. Finally, to allow the roof to roll back, an overlapping hinged board was attached to the south-facing roof wall.

Two layers of heavy-duty green mineral felt were then laid in opposite directions to give the maximum protection from water seepage. It was decided that hot-tarring the felt onto the roof would be overkill, and would further contribute to the weight of the roll-off roof. Therefore $1 \times \frac{1}{2}$ inch (25 × 12.5 mm) battens were applied to the seams. It was now simply a case of sitting back and waiting for the first shower of rain to show

Figure 7.10. Plastic guttering and a down-pipe which directs rainwater into a water-butt adds an eco-friendly aspect to the observatory!

how successful we had been in weather-proofing. As expected we did not have long to wait, and although there were several small leaks, these were quickly sorted out with liberal use of exterior sealant.

The height of the roof is such that it is just possible for me stand up with the roof closed – tall people (over 6 foot) have to adjust their posture accordingly and watch out for the beams! While the roof is quite heavy, once the initial inertia is overcome it rolls surprisingly freely. Obviously, it is important to ensure that the casters are regularly greased. The roof is simply secured into position by a 10-inch (25 cm) hasp and staple on the two side walls. This system has proved effective in all weathers, including storm-force winds.

To help with dark adaptation and reduce general reflections, I decided to paint the interior of the observatory. While not everyone feels this is necessary (many observatories are left with bare wood interiors) I opted for a deep blue woodstain rather than the standard black. I felt this would be visually pleasing while still achieving the desired effect.

I also decided early on that I would lay hard-wearing carpet in the observatory as it would both improve the general insulation from the ground and also act as a cushion should any eyepieces be dropped. On the official opening of the observatory one member of my local astronomical society (the South East Kent Astronomical Society) jokingly commented that the carpet was better than they had in their living room!

While the optical quality is good for a telescope of this nature a number of improvements are needed to bring the instrument up to its peak performance. As the original Meade bands did not allow the telescope tube to rotate, a new set of cradles and slip ring were built. This, with the addition of Teflon sheeting applied to the inner surfaces of the cradles to help reduce friction, has proved reasonably effective and allows the eyepiece to be placed in a convenient position for most observations.

The telescope proved extremely sensitive to improper balance, and I initially had considerable problems with "backlash". However, several hours spent in fine adjustment of the two 40 lb (18 kg) counterweights and worm block did improve things somewhat. While not perfect (there is still some backlash to eliminate) the scope is now very usable. Eventually, I intend to replace the supplied RA drive with a large stepper motor with a maximum useful torque of 200 N cm, and add a further motor to the declination shaft.

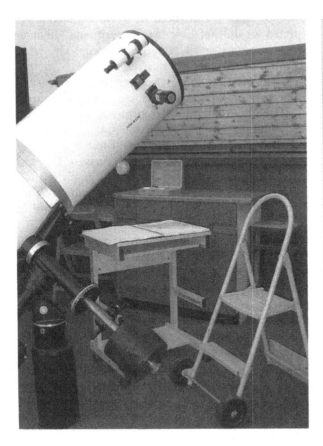

Figure 7.11. A surplus computer table makes an ideal movable desk for star-maps and eyepieces. Note the kitchen unit in the background which acts as general storage space.

Finally, I intend to add a full computer-controlled GOTO capability to the telescope. While I am very much a visual observer, I see computer control as an aid to both enhancing and maximizing your time spent out under the stars. CCD imaging may be something that I will get into in the future, but until that time, the delights of detecting a distant galaxy at the extreme edge of visibility will continue to lure me away from a warm bed.

A permanent observatory is the one thing that truly maximizes your observing time. I can be up and observing literally within a few minutes of checking that the sky is clear. It's simply a matter of unclipping and rolling back the roof, and then swinging the telescope into position! No longer must I spend time in assembling the telescope and then polar-aligning the mount, as I have done for years with smaller telescopes.

Sadly, there are few places left in south-east England with truly dark skies, and while there is some light pollution to the south-west, due to Dover docks, the rest

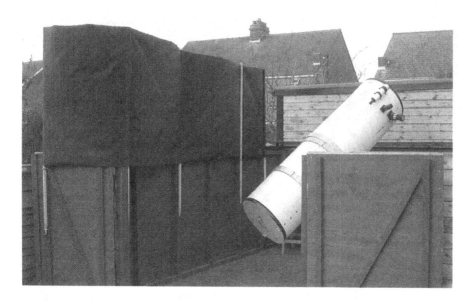

Figure 7.12. A strong
black plastic "sail" can
be quickly erected to act
as a windbreak and/or
to block out any sky-
glow from the south-
west.

of the sky can reach a reasonable visual limiting
magnitude. Interestingly, during winter months I get a
low sky glow from the lights of Calais, France, which is
22 miles (35 km) across the English Channel, and just visible
through some distant defoliated trees in the south-east!

An addition to the basic observatory is a strong black
plastic "sail" that can be erected above the south-
western facing wall to block out the sky glow from the
docks. A further advantage to this system is that it also
acts as an additional wind-break from any prevailing
westerly wind. This has proved to be very effective in
both maintaining dark adaptation, and in general
protection from the elements when I am not observing
objects in that region of the sky.

As time progresses there are bound to be some
modifications to the observatory and telescope. However, as things stand at the moment the set-up seems to
be both effective as an observatory and strong enough
to withstand winter gales and storms, all of which seem
to be on the increase.

If I rebuilt the observatory would I change anything?
In retrospect, I could probably get away with 5-foot
(1.5 m) high walls instead of the current 4-foot (1.2 m)
height with a very minimal reduction in total sky
visibility. This would add a little more protection from
the elements, and would also reduce the depth and
weight of the roll-off roof. Other than that, I am pleased
with the success of the design.

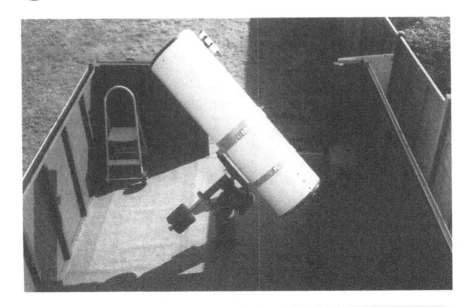

Figure 7.13. This high, wide-angle view gives some indication of the relative size of the telescope and the space within the observatory.

The observatory could not have been built without the invaluable and sustained help of several close friends (Martin, Tony and Paul) and my long-suffering wife, Catherine.

Chapter 8

Ken Dauzat Observatory

Ken Dauzat

Amateur astronomy has interested me from an early age. Raised on a farm with my parents and three brothers, I received my first telescope, a Sears 60 mm refractor, as a gift for my 10th birthday. I will never forget the first time I viewed Jupiter and Saturn with it while sitting on the front porch steps of our old country home. I blushed with excitement as each of my family members took turns looking into the eyepiece with astonishment.

For almost 10 years, after school and on weekends, I worked on my father's 1300-acre farm with machinery and tools. It was hard work but excellent training. After graduating from Marksville High School, I attended Louisiana State University at Alexandria. During my first year of college, my astrophysics instructor offered me his father's old 6-inch, f/8 Criterion Newtonian that rekindled a new flame of interest for me. From that time on, I would be captured by the impressive lunar and planetary images that the old Criterion displayed.

I served in the US army from 1970 to 1972. After being discharged with a permanent foot injury, I still managed to maintain a full-time job with Medical Transportation Services as well as several part-time ones too. I even sold satellite TV systems after working hours and on weekends from 1980 to 1983. Even with all of this, I would still find time for telescopes and astronomy. Being mechanically inclined, I studied telescope designs and was constantly experimenting, dismantling, and reconstructing optical designs and their mountings. Somewhere along the way, I acquired a fully equipped machine shop which I constantly upgrade with new machinery to carry on my work.

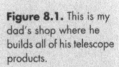

Figure 8.1. This is my dad's shop where he builds all of his telescope products.

After retiring from years of public work and small private business ventures, I now had the time to devote more attention to my greatest hobby. My oldest daughter Angel was married and moved away and this past spring, my son Dwayne graduated from college, in Computer Science and is now working away from home. Our youngest daughter, Ashly has become my only little helper and buddy. The kid is a brain! – with a 4.8 average in Honors and the energy of a small nuclear plant – and I have programmed her with all the carpentry and mechanical skills I know.

During the years from 1980 to the present time, I acquired, restored, modified, and built many interesting telescopes including a 5-inch, f/15, 1889 Alvan Clark Refractor, a $16\frac{1}{4}$-inch, f/12 Cassegrain and mounting, a $12\frac{1}{2}$-inch, f/10 highly baffled Newtonian, and various other telescopes. Several of these were later sold to make room for different models. Eventually I began working on telescopes for friends and building telescope accessories and thus began my latest business, "Ken's Telescope Rings and Accessories". You may access my Web site at: http://users.kricket.net/ken.

It wasn't until about 1998 that I acquired a fork-mounted C11. The optics were very impressive and I decided it would be a keeper. The problems with the older fork-mounted sand-cast models was the weight factor – much too heavy for me. I was now craving with great desire for a permanent observatory of some sort. I consulted Ashly about it and we agreed that treating Moma to a night out would get the bill passed.

Thus began the "Ken Dauzat Observatory". My budget would be about US$1500 and I would have to fit all the frills and fancies into that amount.

After several months of planning I decided to have the observatory near my home for convenience purposes. I found the perfect place, just off the patio in the back-yard. After measuring the available area I decided on a 10-foot diameter round building that would support a 3-meter (10 feet) fiberglass dome. At this time, I would concentrate first on building the slab structure.

I wanted the slab structure to be very solid so that there would be little or no vibration while observing under high magnification. So I decided that the central part of the slab should be much thicker that the outer edge. That would create an extremely solid cone-shaped structure. For the circumference, I began to rip 9-inch (22 cm) strips of $\frac{1}{4}$-inch (6 mm) paneling (damaged sheets purchased from the local lumber yard for US$2 per sheet) to be used on the outer edges of the form. I knew that 32 feet (9.5 m) lengths of these strips would give a form diameter of about 10 feet 122 inches (3.05 m) to be exact. Cutting two dozen wooden anchor wooden stakes, I drove them into the ground on the outer edges at exactly 61 inches (152.5 cm) from the center of the slab. Using a handy screw gun, I screwed the anchored wooden stakes to the paneling. By stretching a nylon string across the top of the form, I dug out the center of the form until it measured 30 inches (75 cm) deep.

For the pier, I would use an 8.5-inch (21 cm) diameter, thick walled, steel pipe that I had left over from the old satellite business. Welding several iron structures at the footing, I placed it level in the center of the form. I then ran an electrical wire inside the pier and connected it to a 110 volt outlet installed near the top side of the pier. This would power the drive and accessories while keeping the inside observing area free of cluttered wires. I welded a $\frac{3}{8}$-inch (9 mm) thick steel cap to the top of the pier, then drilled and threaded it to accommodate the C11/C14 wedge bolt pattern.

I then buried the pier wire and had it emerge at the wall area and up a foot or so to reach another proposed 110 volt wall outlet where it would later be attached to receive electrical power. Next was the electrical wire

Figure 8.2. Top of pier being drilled and threaded to secure the equatorial wedge.

running to my workshop and later to be connected to a small 15 amp breaker. The line would not be fired up until the entire project was completed – several weeks away at this point.

I was now ready for the cement truck and asked for a cement spreader to help finish the slab. Two yards (1.8 m) of cement and a couple hours of work revealed a beautiful flat slab. About this time, while the cement was still setting, I began inserting the 12 steel bolts that would secure the wall structure to the slab. The 5/16 × 6 inch (8 mm × 150 mm) bolts (30 cents each, including the flat washers and nuts) were placed to protrude almost 2 inches (50 mm) out of the slab to allow the 1½-inch (38 mm) thickness of the 2 × 4 inch (50 × 100 mm) timbers later used for the wall base. I was sure to place one on each side of the proposed door area for strength.

A week later I removed the plywood form and wooden stakes that revealed a nice round shape foundation. I was now ready to begin the wall construction. I began by measuring the circumference of the center wall area and discovered that 32 one-foot (30 cm) sections, all angled to match each end, forming a circle, would be required to complete the rounded wall, bearing in mind that 3 feet (90 cm) of the front area would be reserved for the door. After one afternoon of bolting and nailing the base pieces together I was ready to add the 6-inch (150 mm) tall wall studs. Actually, the studs would be 3 inches (75 mm) shorter than 6 feet (1.8 m) to allow the thickness of the top and bottom plates. Next came the stud braces that would give strength to the wall and also allow a solid area to screw on the outer galvanized steel siding.

Figure 8.3. Wooden structure is now drilled and the electrical wiring is installed.

Corrugated galvanized steel tin is an excellent choice for conforming to the rounded wall structure. Ashly came up with the idea of using galvanized steel screws instead of nails. This would create a stronger bond and keep down the pounding vibrations that could loosen or warp the structure. I discovered that, price-wise, sixteen 6-foot (1.8 m) sheets of tin were much more expensive that eight 12-foot (3.6 m) lengths and cutting them in half would work.

Figure 8.4. With the siding and door completed the dome track and bearings are now installed and tested.

After the siding was complete, I began thinking about the door construction. I chose to go with 1 × 6 inch (25 × 150 mm) decking material since it had rounded edges and was treated for rot resistance. I measured the door-way opening and went to work bolting the door structure together. Heavy galvanized steel hinges were used and a large door latch was bent in my shop vice to fit the wall and door closure. After several weeks the door would be removed and finished in an antique blue-green color.

Now the time had come for adding the steel rim track that would support the rotating dome. After thinking about this cautiously, I decided to use 2 × ⅛-inch (50 × 3 mm) steel flat bar for the rail. This would rest on eight 3-inch (75 mm) grooved steel pulleys with center ball-bearings. How much would these special pulleys cost? Our Alexandria Lowe's lumber yard offered them for clothes lines for only US$2.28 each!

We now decided to take a break and think about the most difficult part of all – the Dome. At this point, the budget was still intact and all was well. Even Moma (Karen) approved of our work so far.

After a week or so thinking about it, Ashly and I decided to tackle the dome. First would be the base ring. We would use a double layer of ¾-inch (18 mm) plywood glued and then screwed together with dry wall screws to create a solid 1½-inch (38 mm) thick base. This would be the backbone of the structure on which the ribs would rest.

Figure 8.5. The dome base being constructed on a hard flat surface.

To achieve a perfect cut circle for the base and ribs, we devised a simple method of using a 1 × $\frac{1}{8}$-inch (25 × 3 mm) thick steel flat bar bolted to the bottom of the old skill saw and drilling a guide hole at the center of the circle. This turned out to be about 60 inches (150 cm) from the saw blade. At that point Ashly embedded a dry wall screw into the pilot hole at the end of the flat bar and on the center line that we had drawn down the middle of the $\frac{3}{4}$-inch (18 mm) plywood. Choosing to have the ribs about 4 inches (100 mm) wide, we had also to drill a second pilot hole up 4 inches toward the saw for the inside cut of the rib. I would like to note that the saw blade must have quite a bit of setting in the teeth before it will clear the curve that you are attempting to cut. To achieve this, simply remove the blade and clamp it in a vice and, using an adjustable wrench or Crescent Wrench as some would call it, bend the teeth of the blade outward a little on each side. This will allow the blade to cut a wider path and thus clear the circle. It is advisable that you use an old blade as this will probably ruin it for any other use.

Figure 8.6. The skeleton dome frame is assembled using glue and drywall screws.

We estimated the number of ribs we needed and spent the entire afternoon just cutting them. The next day we began assembling them, beginning with the opening sections first. One question in our minds was how wide should the observing opening be? That

Figure 8.7. The thin plywood panels are now secured to the skeleton frame.

depends on the telescope aperture that you will be using. Too wide would allow dew and moisture to settle onto the optics, and too little would not allow enough observing time before the dome required turning as the telescope tracked the object you are observing. I decided that near 24 inches (60 cm) would be sufficient even if I later replaced the C11 with a C14, which I eventually did! We then painted the entire rib structure with a navy-blue latex.

Next came the addition of the $\frac{1}{4}$-inch (6 mm) exterior grade plywood shell. Again, dry wall screws were used to attach the plywood to the ribs. Ashly would mark the cuts inside the dome while I would curve the 4 × 8 foot (1.2 × 2.4 m) sheets around the rib openings. Then the cuts would be made 1 inch (25.4 mm) wider that the pencil marks to allow enough area to screw onto the ribs. Remember, the ribs were limited to create a $1\frac{1}{2}$-inch (38 mm) thickness. When all was complete, the panels were removed and painted flat black, front and rear, then allowed to dry two days before being replaced.

It was now time to place the dome on top of the wall structure. We accomplished this by recruiting several neighbor helpers and, placing them around the dome, lifting it over the top of the walls and letting it down onto the rail support. The side skirting was added using a 10-inch (25 cm) wide, $\frac{1}{4}$-inch (6 mm) paneling bolted to the base ring. Then a $1\frac{1}{2} \times \frac{1}{8}$-inch (38 × 3 mm) steel

Figure 8.8. Dan is spreading a fiberglass blanket and Darian and I apply rosin over it.

flat bar was riveted on to the inside of the skirting to reinforce the edge. We were now ready for the fiberglass coating that would seal the outer section of the dome and make it waterproof.

Now was the time to call on my old friend who owned a fiberglass business. Dan was an expert in the fiberglass business. He offered to help me for free, provided I pay only for the material. I insisted on paying him but he would hear nothing of it. My best observing buddy, Darian, also came over and assisted. Dan began by stapling the 24-inch (60 cm) wide fiberglass cloth across the dome top and pouring the rosin, spreading it with a paint brush as he went along. Darian and I copied his technique "somewhat" at the lower areas until we reached the bottom edges. Together we worked from 9:00 that morning until 4:00 that afternoon and finally completed the job.

I wanted to add a motorized electrical device to rotate the dome and checked out several methods. After thinking about it carefully, I then decided on using a roller chain, tightened around the skirting to act as a huge gear. A #40 motorcycle chain would work and offer great strength. At this point the dome would rotate very easily by hand. I aligned the chain by rotating the dome and drawing a complete circle with a felt marker in a fixed position and then tapping the chain down to the edge of the line. Rosin was then applied over and under the chain to lock into a fixed position and formed an integrated look with the dome. Two days later Ashly and I applied a heavy coat of beige color enamel with a roller and extension, finishing the edges with a small brush. One gallon was sufficient to cover it thoroughly, including the roller chain.

Now was the time for the dome cover to be installed. I purchased a piece of $\frac{1}{8}$-inch (3 mm) thick aluminum, 30 inches (75 cm) wide and 96 inches (2.4 m) long and had it rolled at a local machine shop to the contour of one of the leftover plywood ribs. The total cost of the cover was $75, which I though was cheap enough. Using eight sliding glass-door grooved rollers with ball-bearings (US$1.29 each at the hardware store), I drilled holes and bolted them to the upper and lower edges of the aluminum door. They would then ride on a $1 \times \frac{1}{8}$ inch (25 × 3 mm) aluminum flat bar, used as rails and bolted to the top and bottom of the opening frame.

I now needed a drive motor to power the dome. I acquired a 48 volt geared motor with a matching #40 ten-tooth gear welded to it. These motors can be purchased from several suppliers on the Internet and sell for under US$100. Search under "gear head motors".

The motor was then secured to the wall structure using 3-inch (75 mm) long lag bolts into the 2 × 4 inch (50 × 100 mm) timbers. It is wise to pre-drill a slightly smaller pilot hole before tightening down the lag bolts. This will prevent the 2 × 4s from splitting. I placed a small backup bearing inside the skirting to keep the roller chain in constant mesh with the geared motor. To power the motor I used two 24 volt DC transformers connected to two, three-way switches to allow 24 volt or 48 volt

Figure 8.9. Detail of the geared motor in mesh with the dome rotating chain.

Figure 8.10. The inside paneling is completed and the telescope is mounted onto the pier.

operation for two speeds and in clockwise or counter-clockwise rotation. A large aluminum plate covered the opening where the transformers were connected to 110 volt current. This access also allows for easy replacement of the transformers, should they ever need it.

To complete the inside electrical system, I added three more wall outlets (you can never have enough of these) and two 40-watt red lights in recessed sockets installed with a zoom dimmer switch. Then the paneling was nailed to the inside. $\frac{1}{8}$-inch (3 mm) solid wood paneling was chosen to allow easy bending around the curved interior. At the lumber yard, I found several sheets that had damaged edges that would work since only 6 feet (1.8 m) of the sheet would be used. The cost for these was less that US$20.

It was now time to install the C11 wedge and finally the telescope itself. Moldings and wall pictures completed the interior trim. The furniture consisted of an ocular and chart desk, two easy chairs and a bar stool for observing. The total cost for all the materials

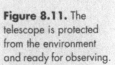

Figure 8.11. The telescope is protected from the environment and ready for observing.

and slab was under US$1300. The leftover funds went to the Ashly (my daughter) foundation.

Almost a year later I replaced the C11 with a fine fork-mounted C14 that I completely refurbished with the addition of tube weights, handles, AP type focuser, and additional guide-scopes.

After a hard day's work and a hot bath, I can now sneak out of the back door and be observing within five minutes, right from my patio.

Figure 8.12. The completed Ken Dauzat Observatory.

Chapter 9
A Lancashire Observatory – Part II

David Ratledge

Introduction

In the first book, *Small Astronomical Observatories*, I described the construction of my glass-fibre observatory, but of course that was only half of the story. An observatory is only as good as the equipment it contains and it is appropriate therefore in this, the second volume, that I complete the story by describing how an empty shell became a functioning amateur observatory. But first I must thank my co-workers who, with myself, make up the Bolton Group of Telescope Makers, namely Gerald Bramall and Brian Webber. The three of us are from that generation where making telescopes was the only option and without the combined efforts of all three of us a project such as this would have been a daunting task.

Readers of the first book will recall that I chose a classical dome largely as this offered the best protection from neighbourhood lighting which plagues my location. The design of my telescope would similarly require careful thought if it was to minimize the effects of light pollution. Our mission was to construct a state-of the-art telescope for CCD imaging, namely:

- Lightweight tube assembly
  Optimized baffled design to eliminate stray light
  Electric focuser for precise focusing
- 400-mm/16-inch aperture
  Fast focal ratio for quick imaging (f/4.7)
  Precision optics – 1/30 of a wave or better

Figure 9.1.
Observatory as
completed in *Small
Astronomical
Observatories.*

- Computer-controlled pointing and tracking
  GO TO invisible objects
  Accurate tracking with Periodic Error Correction
  (PEC) for unguided exposures.

The tube assembly consists of an aluminium frame
constructed by Gerald, the optics by Brian and the
mount by myself. Our working name for the telescope
was Stealth, which came from our attempt to cut out
light reflections. A tube design using diaphragms was
adopted rather than the more usual Serrurier truss type,
which is more suitable for dark skies. Diaphragms or
baffles provide the best weapon against stray light by
preventing it from bouncing off the inside of the tube
and reaching the focus by devious routes. The central
aperture in each baffle increases away from the mirror
and matches the field of view of the telescope. In this
design the baffles also act as structural elements which,
together with six longitudinal 50 mm (2 inch) diameter
aluminium tubes, frame it all together. Clearly visible in
Figure 9.2 is the structure of the telescope but on the
finished telescope (Figure 9.5 below), plastic panels
have enclosed the tube, keeping out stray light but
masking the construction.

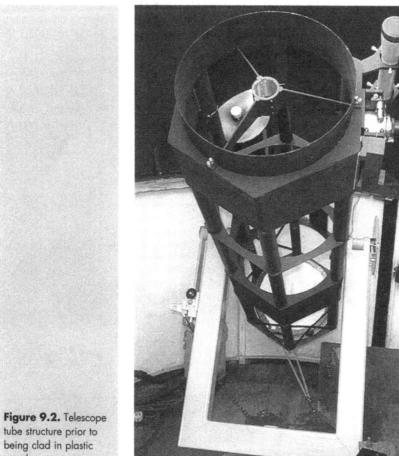

Figure 9.2. Telescope tube structure prior to being clad in plastic panels.

The Tube Assembly

The tube comprises six hexagonal baffles and six 50 mm diameter aluminium tubes. The tubes are not continuous but are in fact in sections. Through the tubes pass full-length threaded rods which, with locking nuts at each diaphragm, clamp the whole structure together. This creates a strong but very light structure. Diagonal cross-bracing was added to the lowest section, i.e. next to the main mirror, to stiffen this the heaviest loaded part. Aluminium plates were added between the diaphragms at the point where the declination shafts are attached and also one at the top for the focuser.

The mirror is supported on an 18-point suspension system made entirely of aluminium, which is very low profile, keeping the tube as short as possible (Figure 9.3). The geometry of its arrangement was calculated using the *Sky & Telescope* Basic program which is available

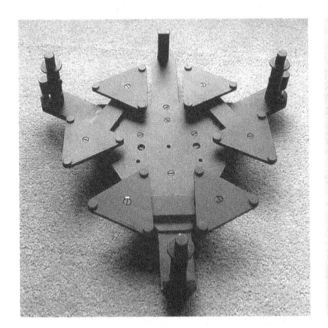

Figure 9.3.
Low–profile mirror cell

Figure 9.4. The 80 mm diameter Crayford focuser carrying the CCD camera.

from their Web site (www.skypub.com), although finding a computer to run Basic is not easy these days. Adjusting collimation is via three big knobs and it is very smooth and straightforward. The finder was made from Russian 20 × 60 binoculars, probably the cheapest source of optics this size. A Telrad is also used.

To carry the large weight of the CCD camera a heavy-duty 80 mm diameter Crayford focuser was constructed (Figure 9.4). Crayfords are by far the best focuser and although now available commercially (at a price) they are comparatively easy to make. The 80 mm tube was sourced from an old photocopier drum and the four ball-bearings that it rides on are from scrap 3.5-inch floppy disk drives. It was motorized with a 12 volt instrument type electric motor, which provides precise movement, one quick push on the button moving the focuser just 1/1000 inch. The Crayford has no trouble holding the weight of the CCD camera even when a filter wheel and a ×3 converter are both in line with it.

After initial testing, the open skeleton tube was totally clad with 3-mm ($\frac{1}{8}$ inch) thick lightweight plastic, similar to plasticard available from model shops. The panels were stuck on with proprietary adhesive, the type that is sold under a variety of names such as "No Nails", "Liquid Nails", etc. Three panels, which are removable for access, are held in place with Velcro. The plastic cladding has provided the protection of a solid tube but without the weight. It has been painted blue, resulting in a name change for the telescope to Blue Streak. I hope it is more successful than its namesake (for those not familiar with Britain's failed attempt to join the space race, Blue Streak was a rocket which invariably crashed!).

The Optics

The mirror was made by Brian on his home-made grinding machine. It is made from 40 mm thick Pyrex sheet and has a focal ratio of f/4.7. Pyrex mirror blanks are very much harder to find these days and this one had to be cut from a 20-inch (500-mm) square piece – in other words a 20-inch mirror blank would have cost the same! The tool used for coarse grinding to the correct profile was made of steel. For fine grinding this was covered with small glass squares, stuck down with proprietary gun adhesive ("Liquid Nails").

Figure 9.5. Finished telescope on the fork mount.

Once fine grinding was completed then the polishing stage could begin. It was polished with a subdiameter tool and taken to a sphere first. This is the easiest surface to make and test using a Foucault tester. The Foucault tester was specially modified to get the source (usually a pinhole) and the knife edge as close together as possible for this fast focal ratio. This was achieved by using a fibre-optic instead of a pinhole, which has the added advantage of producing a much brighter image.

Parabolizing was the final stage and progress was assessed by means of the Dall null test. This simple test uses a lens to introduce equal and opposite spherical aberration to that of a parabola and hence the mirror is checked for a simple null, as if it were a sphere. Once this test exhibited straight ronchi patterns then zonal readings could be taken. From these the mirror's accuracy was calculated using the formulae derived by Texereau. The wavefront error in the final mirror is around 1/30th i.e. 1/60th on the actual glass – it shows textbook ronchi patterns before, at and beyond focus. The diagonal mirror is 3.5 inches (89 mm) minor axis and was one of the last made by Hinds Optics (UK).

Computer Controlled Mount

The mount was built using welded steel, 150 × 100 mm (6 × 4 inch) rectangular hollow section (RHS) for the support and 100 mm (4 inch) square RHS for the fork. The polar axis is 2-inch (51 mm) solid steel and rides in two self-aligning ball-bearings with the thrust taken on a 25 mm (1 inch) steel ball-bearing at the bottom end of the shaft. For computer control it has to be driven in both axes (RA and Dec). A sector arm, which is commonly used on the Dec axis, is not suitable. The RA drive has a 10-inch (254 mm) diameter Matthis worm and wheel whilst the Dec has a Byers 7.5-inch one.

What is meant by computer control? Well, I mean being able to control the telescope's operation, position and tracking by means of a computer. In other words the computer will "know" where the telescope is pointing and, when commanded to move to an object, will be able to do so with precision. Having got there it will be able to correctly track the object, be it a galaxy, planet or comet.

Over the years various solutions have been adopted to bring a telescope under computer control but for our purposes the options boil down to stepper motors or servo-motors with encoders. The former know position by counting the "steps" from a known starting position (i.e. a star) whilst the latter use encoders to determine position. The steppers are probably the easiest for the amateur with the servo-motor drive more suitable for advanced systems. Nonetheless, as we shall see, when equipped with sophisticated driver software, the simple stepper is capable of excellent tracking, slewing and, most importantly, high-precision pointing.

It is obviously possible to computer control alt-azimuth mounted telescopes (see Mel Bartels' Web site, www.efn. org/~mbartels) but it is probably more practical to use an

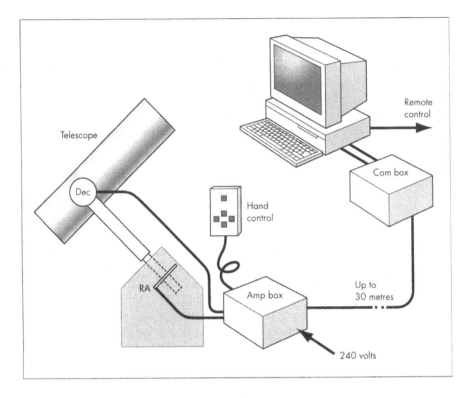

Figure 9.6. Schematic layout of the control system. The Combox and PC can either be located in the observatory or remotely indoors.

equatorial mount. Not only does it make life easier (no field rotation to worry about) but it gives us the possibility of piggybacking other telescopes and cameras on the mount. They would also require field de-rotation on an alt-azimuth mount. Another disadvantage of driven alt-azimuth telescopes is that they have a dead zone around the zenith. But this is the best place to carry out observation, as light pollution and atmospheric absorption are at a minimum there. So an equatorial is easiest but because the telescope will be moved under computer control both axes have to have full drives with stepper motors as opposed to traditional synchronous motors. We will need drivers for them and power amplifiers. To control all this a dedicated computer (an old 486 is fine) with control software is required. To summarize, we need:

- an equatorial mount with full drives to both axes
- two stepper motors
- stepper drivers with amplifiers
- an old PC
- an ISA bus counter card
- controlling software.

Figure 9.7. Stepper motor and RA gears.

Whilst dedicated computer programmers could no doubt write their own control software I decided to use commercially available software. The system I chose was PC-TCS by Comsoft, which is used worldwide on a variety of small professional (i.e. up to 1.2 metres) telescopes. The owner of Comsoft, Dave Harvey, says he prefers to deal with amateur astronomers as he has less trouble with us than professionals! It can be bought as a complete system ready and working but I chose the (cheaper) "kit of parts" option, i.e. software plus all the components loose. This duly arrived and whilst it was daunting at first, tackled in easy stages, it was relatively straightforward to assemble.

The first job was to install the new stepper motors, which because of their size was a major task. They are much larger than conventional motors as they have to be capable of slewing the telescope in any direction and at speed. The Dec incorporated a 7.6-inch high-precision Byers gear set bought cheaply through Astromart. The RA also needed a strong housing and Brian machined a worm housing with extended shaft, again to provide clearance for the gears and motor (Figure 9.7). Both gears systems are covered for safety.

The goal on the mechanical side is to reduce backlash by as much as possible. If you can, go for a worm and wheel with sufficient teeth so as not to need spur gears. I needed 2:1 spur gears which is less than ideal as they are bound to introduce some backlash, which will reduce pointing accuracy. It is best to disable any clutches in the drives as these too can be a source of backlash or even slippage.

With the mechanical side more or less sorted, attention turned to the electronics. A basic DOS PC was obtained with just over 600 kb of free base memory – essential for the TCS software. The timing card, with its daughter take-off, were installed and these provide connectors for the RA and Dec. The combox, which takes the inputs and outputs from the card in the computer, was next. This was really only housing the circuit board and providing a home for the output connector. An old Ethernet router box was used.

Next job was the amplifier box which contains the two micro-stepping amplifiers, power transformer plus all the connections from the combox and the hand-paddle (Figure 9.8). I obtained another old Ethernet repeater box (bigger than the combox) and the parts were installed into it. This was the biggest job and involves careful wiring and soldering.

The hand-control paddle was built using standard components, available from any good electronics supplier. It is essential for moving the telescope – remember once under computer control you cannot manually move the telescope or it will get confused over its position, which can be dangerous! The hand-control box has momentary-make switches for E, W, N, S, plus a guide/drift switch and a fast slew button. The wiring of the connectors from the combox to the hand-

Figure 9.8. The ampbox contains the transformer and two amplifiers.

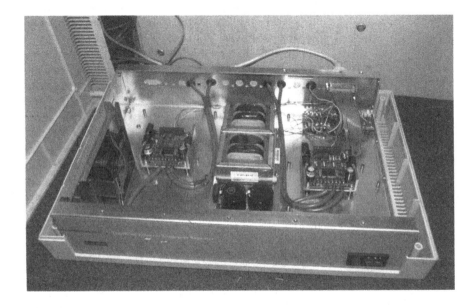

controller was the trickiest part. The connections for the transformer and two micro-stepping amplifiers was fairly straightforward.

For initial testing I took the motors off the telescope and drove them unconnected. The thought of driving a $\frac{1}{4}$-ton telescope around without first passing my driving test was too frightening! Once completed, switch-on was a disappointment – nothing happened – just a hum from the transformers and an error message on the screen! It took a few emails to Dave Harvey at Comsoft to finally track down that the RA and Dec cables were transposed. Once this was sorted, PC-TCS sprung into action! When initially started up the telescope is assumed to be pointing to the zenith and stationary. Commanding tracking to start, the RA motor began turning. I then selected the next object; the Sun would do. Then by issuing the Move>next command, both motors sprang into life accelerating up to full speed. The displayed telescope coordinates changed rapidly as

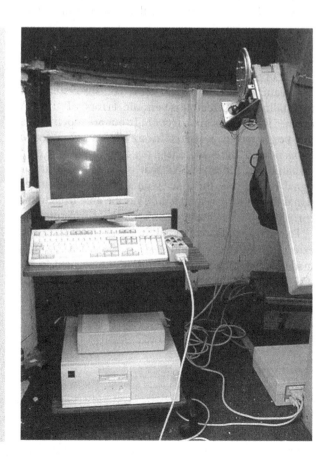

Figure 9.9. Telescope Control System (TCS) installed in the observatory – cables everywhere! Note the Dec worm drive, before the fitting of a protective cover.

they zeroed in on the Sun's coordinates, slowing down as they approached and finally locking on. The hand paddle was checked next – yes all the direction buttons produced the appropriate response from the motors. It was looking good!

When it was coupled up to the telescope there were remarkably few teething problems to sort out. The motors ran the wrong way initially so a couple of wires needed reversing. The system has to be started up with the telescope pointing at the zenith – this "stow" position was easily changed later.

For the first test, I soon discovered when slewing a $\frac{1}{4}$-ton telescope around at high speed that the clutches on both axes needed considerable tightening up – otherwise there was slippage with the system losing pointing accuracy. Once they were tightened up the "Go To" worked well although backlash still needs reducing. Tracking was switched on and the main worm turned slowly – with my gearing there are over six micro-steps per arc second, so the movement is smooth. The visible horizon was entered so that the control system knows not to point at houses and trees. The telescope acquired its first invisible object at the end of August 1998 – the cluster NGC6791 – and although not dead in the middle of the CCD it wasn't far off!

A feature of PC-TCS (and most control systems) is the ability to correct for periodic errors of the RA worm. These errors are a fact of life but are repetitive so can be corrected by the software "learning" a set of corrections. These are then replayed back for each revolution of the worm. All that is needed is an index signal to indicate the start point. Without this index signal the control system would not be able to play the corrections in synchronization with the worm rotation.

To provide the index signal the main worm shaft was extended and on it was mounted a disk with a "bump", made by fastening a washer to it. Mounted next to the disk is a switch with an arm and mini-roller which bears on the disk edge. When it reaches the "bump" it is pushed up depressing the switch. This signal provides TCS with the zero point.

Conclusion

The observatory and telescope have now been operational for two years and have achieved or exceeded all

Figure 9.10.
Completed observatory
and computer controlled
telescope.

the design goals I set out. The observatory can routinely
find and place on the chip of my CCD any commanded
target. From my light-polluted site these targets are
invariably invisible and yet I can locate them without
any trouble. The drive system is accurate enough for
unguided exposures of 30 seconds (at a focal length of
75 inches), which, with my fast focal ratio, is sufficient
to produce a definite image. Control software enables
me to take a sequence of images and co-add them,
producing the equivalent of a long exposure. I still have
to go to the observatory to set an exposure sequence
going and perhaps one day I could bring the control
computer indoors. But by actually being at the
observatory I can watch the sky during exposures and
as a result I have viewed many meteors and, on
occasions, aurorae, sights which I would have missed
had I been indoors.

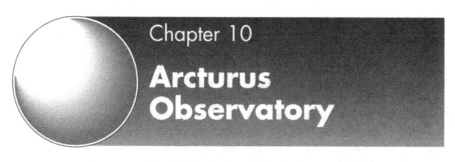

Chapter 10

Arcturus Observatory

Paul Gitto

I would like to dedicate this chapter to Dr. Erik Zimmermann, who has dedicated over twenty-five years of service to ASTRA, The Astronomical Society of the Toms River Area. He has just retired as a professional astronomer, and is now planning to build his very own small observatory.

Figure 10.1. The Arcturus Observatory

The Arcturus Observatory is an astronomical observatory located in the Pine Barrens of Whiting, New Jersey, USA. The Arcturus Observatory was constructed in 1996. The main focus of the observatory is CCD imaging and astrophotography. The Arcturus Observa-

tory is owned and operated by Paul Gitto, DDS. Most of the images are taken with the Meade Pictor 416 CCD Camera. Many of the images can be viewed at the observatory's Web site. The domain name is <u>Comet-Man.com</u>. Images at the Web site are copyright of the Arcturus Observatory, and are for personal use only.

Our observatory is a "Pro Dome 10". It is a 10-foot diameter fiberglass dome 8 feet high. Technical Innovations, Inc. fabricated the dome. Our observatory contains an f/10, 10-inch Meade LX200 telescope. A Meade Pictor 416-XTE CCD camera is currently used. Most images are taken at f/3.3, when an Optec MAXfield 0.33 focal reducer is added. A computer in our home controls the telescope and camera. The dome is yet to be remote controlled or motorized.

When asked to write a chapter for this book, *More Small Astronomical Observatories*, I realized that it could be considered a "large observatory with a very small footprint." Some members of the local astronomy club call it the Mini Keck. Recent advances in technology have made it cost-effective for amateur astronomers to own their own large observatory with a very small footprint. The advances in computers, CCD cameras, and computer-controlled telescopes have brought serious astronomy to an affordable level. Personal astronomy has taken a giant leap forward in the last few years. With the advent of affordable, computerized telescopes, astronomical CCD cameras, sky and image processing software, and the wealth of astronomical data on the World Wide Web, the world of astronomy has opened up to a universe of possibilities. Individuals can now accomplish what was once only possible at only a few of the largest observatories in the world. The computerization of the telescope, combined with detailed accurate software, has eliminated much sky-hopping, and sky-chart navigating. A celestial object can be accurately located within seconds. When it comes to light-gathering abilities, the CCD camera makes up in sensitivity and time, what much larger telescopes had in size. Of course, these observatories are using CCD cameras as well. Image processing software can unblur an image, bring out the faintest details, and even eliminate most light pollution. It is even possible to image deep sky objects on full moon nights! The introduction of the RealSky CD-ROM, a set of CD-ROMS of complete images of the sky, has made it possible to bring actual images of the sky on to a computer screen. These

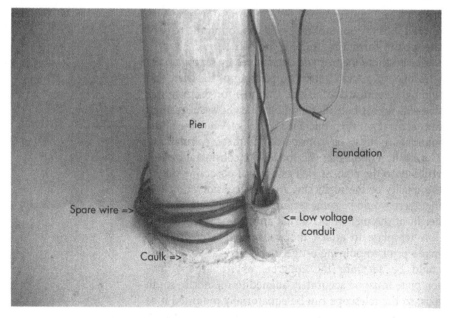

Pier

Foundation

Spare wire =>

<= Low voltage
conduit

Caulk =>

Figure 10.2. A solid pier is necessary for any observatory. Note that the pier is separate from the foundation, and the space is filled with caulk.

images can be compared with the image just taken by the CCD camera. Asteroids can be hunted down, supernovae found, and comets discovered. In addition, the power of the Internet has been unleashed. We can upload and download images, as well as learn from each other how to improve our techniques. This is the golden age of amateur astronomy.

But let's get back to the task at hand, which is how to build a state-of-the-art "large observatory with a very small footprint". The main components of a small observatory are the pier, the foundation, electric wiring, the dome, the telescope, the camera and the computer with software.

The Pier

The pier is the backbone of the observatory, and until it is satisfactory, additional construction should not continue. Our pier was constructed with concrete and steel rebar. The form was made from Sonotube. Sonotube is a cardboard tube into which the concrete is poured. To achieve the desired length and width, two 4-foot (1.2 m) long Sonotubes of 8-inch (20 cm) diameter were connected with duct tape. This made an 8-foot (2.4 m) long column of which over 5 feet (1.5 m) were placed underground for support. Concrete was poured into the Sonotube, as well as around the base of the tube to form a footing for the pier.

The height of the pier was determined by setting up the telescope on a tripod, prior to pouring of the pier, and finding a comfortable height. The hole for the pier was dug with a post-hole digger and the height was measured for accuracy. The thickness and elevation of the foundation must be considered in these calculations. The location of the pier was made with considerations for visibility to the pole, as well as optimal observing due to the quality of the night sky. Another consideration was that there would be some settling of the concrete, and the height of the pier would shrink an inch or so.

A template to attach the wedge and telescope was made prior to pouring of the pier. This was so anchors could be set into the concrete as it hardens. The template must be accurately aligned to the north–south axis, so the telescope can be equatorially mounted if so desired. It is important to remember that true north is not magnetic north, and accuracy of template placement is highly important. The pier at the Arcturus Observatory had to be redug, and repositioned due to the fact that the steel rebar in the pier had acted as a magnet and skewed the measurements just enough so that polar alignment was not possible. So to reiterate, remember prior to pouring the foundation, it is extremely important to test the pier for polar alignment and make sure it is level. It would not be a pleasurable job to have to remove a new foundation. We were fortunate to have tested our pier prior to pouring the foundation.

The Foundation

The foundation of our observatory consisted of a 5-inch (125 mm) thick steel reinforced concrete slab sitting on a bed of crushed stone. The foundation also has a $3\frac{1}{4}$-inch (8 cm) sponge spacer so that no vibrations are passed to the pier. A concrete walkway leads to the observatory. This walkway is very helpful during the winter months. A quick shoveling prevents snow getting carried into the structure. Prior to pouring a concrete foundation, for the observatory, many things needed to be considered.

1. The location of the pier within the observatory. The determination of the position of the pier within the

observatory was made prior to the pouring of the foundation. We placed our pier at the center. Although this works fine, placing the pier south by one foot would have given more workspace on the north side of the observatory. The telescope, which was equatorially mounted, would be better positioned further south. However, a centered pier can help turn the foundation into a gazebo with a centered table, should the observatory need to be removed in the future.

2. The placement of the observatory upon the foundation. We used a circular foundation 2 feet (60 cm) in diameter larger than the observatory. This helps with yard care around the observatory. An additional foot or two would have been a nice luxury, as it would have made walking around the observatory easier.

3. The elevation of the observatory. The foundation should be set a few inches above ground level to allow for proper drainage.

4. The material used: should the floor be wood or concrete? Wood cools faster, and is more forgiving. It is easier to correct mistakes. Concrete gives a more solid foundation, but retains heat.

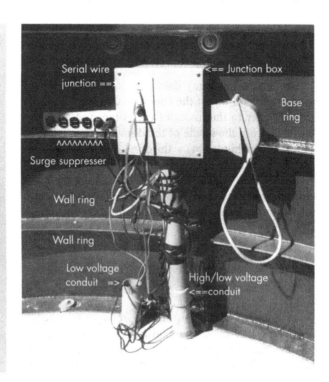

Figure 10.3. The junction box connects the wiring from the house to the equipment in the observatory. Only low voltage runs to the pier.

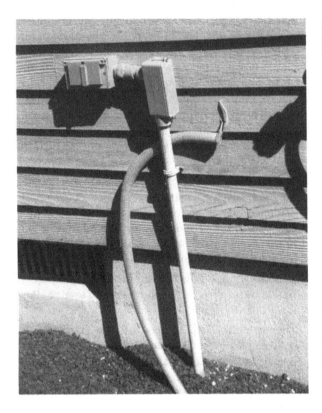

Figure 10.4. A second flexible conduit was added to keep low- and high-voltage wires separate.

5. The location of conduits to run wires from the pier and to the electricity source, and to the computer.

Two 2-inch (50 mm) diameter conduits were laid prior to the pouring of the concrete. One conduit leads from the pier to the inside of the dome, and a second one leads from the inside of the dome to the outside of the dome. In hindsight, a third conduit leading from inside the dome to the outside would have been highly advisable. This would give separate conduits for high- and low-voltage wires. A flexible tube was inserted into our second conduit to keep the high- and low-voltage wires apart.

Wiring

A 120 V line and two low-voltage "copper telephone" lines were run from the house to the observatory's junction box. Separate low-voltage lines were run from the pier to the junction box. These included power and

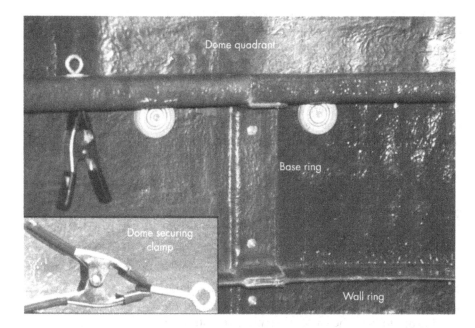

Figure 10.5. The dome rides upon rubber wheels attached to the support ring. Four screws and clamps secure the dome when not in use.

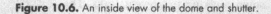

Figure 10.6. An inside view of the dome and shutter.

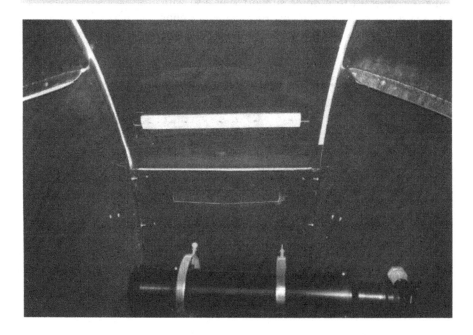

communication lines for the telescope and CCD camera. There was no need to run high-voltage lines to the pier. As a safety precaution, all cables are disconnected from electronic equipment when not in use. This helps prevent electrical storms from shorting out the equipment. Additional low-voltage wires were added later, as it seems there are never enough wires. A word of caution! When reconnecting the wires to telescopes, cameras, and computers it is highly advisable to have all equipment shut off during the connections. It is easy to short out the circuit boards and chips.

The Observatory Dome and Base

Part of the wall of the observatory could have been built with wood, and the dome and base section placed on top. I chose to have the entire observatory constructed with the same fiberglass. There are a few main components to the structure. The dome quadrants and shutter assembly, dome support ring, base ring, and wall rings, and door assembly. The four dome quadrants and one shutter are connected with nuts and bolts, as well as some caulking to prevent leaks. The dome is connected to a dome support ring also with nuts and bolts. The dome support ring contains a door (Figure 10.7). The door carries through to door segments in the base and wall rings.

Figure 10.7. The door secures the dome, and the dome should never be rotated with the door open.

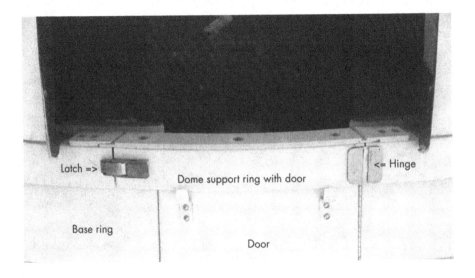

Latch =>   Dome support ring with door   <= Hinge

Base ring

Door

This dome structure sits and rotates upon the base ring. It turns easily on hard rubber, ball-bearing rollers, 3-inch (75 mm) diameter, mounted in the top of the base ring, so easily in fact, a child can move it. The base ring also has a door, which aligns with the dome support ring. The wall rings add height by one foot each. The Arcturus Observatory has two wall rings, adding 2 feet (60 cm) in height to the observatory. There is plenty of headroom with two wall rings. However the height provided by the two wall rings makes it quite difficult to climb in or out, without opening the door. In hindsight, I would have preferred to lower the observatory by one foot (30 cm), and be able to climb in and out without having to open the door each time. The dome must never be moved when the door is open, or the dome can fall off the base ring as it rotates. When the observatory is not being used, and as a safety precaution with many people around, the dome is secured to the base ring with eye bolts and spring clamps. Instead of securing the bolts with nuts, spring clamps are used. They hold sufficiently, are much easier to remove, and don't get lost in the dark!

Figure 10.8. The observatory is secured to the foundation by screws into concrete anchors. The anchors were placed after the foundation was cured.

The wall ring is attached to the foundation with concrete anchors. These anchors are drilled into the concrete after the rings are pre-assembled. During construction, it is extremely important that the observatory is circular. If it is not circular the dome

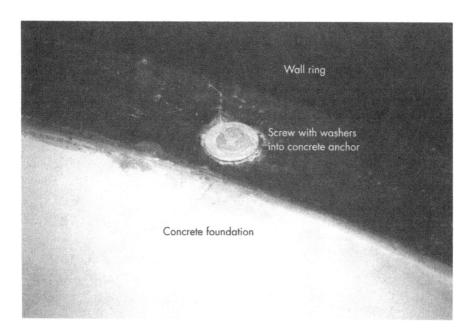

Wall ring

Screw with washers into concrete anchor

Concrete foundation

will fail to rotate. Small intolerances are allowed in the wall rings, but the base and dome rings need to be almost perfect. Technical Innovations, Inc. offers to pre-assemble the observatory prior to delivery. This is a wonderful service, and simplifies installation. It becomes a simple matter of connecting the parts.

Observatory Addendum

Technical Innovations, Inc. was a great company to work along with. John and Meg Menke run a nice mom and pop business. Technical Innovations provides an excellent manual for construction and use of the dome. The manual is very complete and needs to be followed in order for the dome to operate properly. The Pro-Dome has given over four years of excellent performance, and requires very little maintenance. On the outside, I keep it clean by washing it with soap and water; and applying a coat of car wax. On the inside, I keep it clean by sweeping, and dusting. I also occasionally check the nuts and bolts for tightness.

The beauty of fiberglass is that it can easily be repaired. I learned this the second day I had the observatory in operation. Unfamiliar with all the

Figure 10.9. The view of a repaired shutter shows that fiberglass is a wonderful material for an observatory. It holds up to abuse, and can be repaired quite easily.

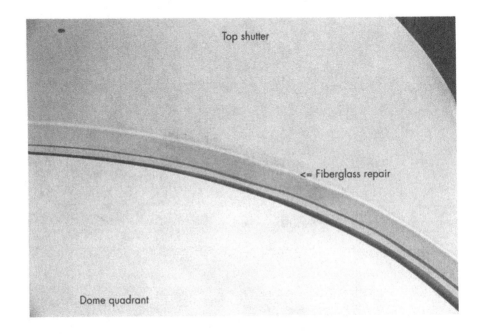

Top shutter

<= Fiberglass repair

Dome quadrant

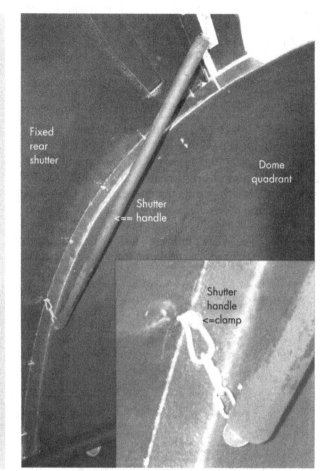

Fixed
rear
shutter

Dome
quadrant

Shutter
<== handle

Shutter
handle
<=clamp

Figure 10.10. This is a simple way to open and close the shutter.

nuances of the shutter operation, I managed to dislodge the top shutter, crashing it to the ground. Aside from a loud noise, a cracked shutter, and damaged pride, there was no major damage. I purchased an epoxy fiberglass repair kit, at a local boating store, and learned why fiberglass is such a wonderful product. However, handling fiberglass improperly can result in skin injury.

Due to the 8-foot (2.4 m) height of the observatory, I had to design a shutter door handle that would be easy to reach and operate. At the local Home Depot, I found a 4-foot (1.2 m) section of steel pipe, some chain, pipe insulation and clamps and made a permanent retractable handle. After looping the chain through the pipe and shutter's main handle, I secured the chain and put a clamp on the dome also. The handle stays comfortably near the dome and does not hit anyone in the head. For safety and comfort, rubber pipe insulation was added. This is great during the winter months as a metal pipe can be quite cold.

The Telescope

The Meade 10-inch LX 200 f/10 is the workhorse of our observatory. It is used with an Optec 0.33 focal reducer, which maximizes the telescope's CCD capabilities at f/3.3. It had always been reliable, and with its GOTO capabilities, it always puts the image on our CCD chip. The telescope is remotely controlled and permanently polar mounted. This makes it a great small observatory telescope. The only limiting factor is that with accessories attached, I keep away from the polar area,

Figure 10.11. The telescope is ready to go, and has a wide-field 4-inch refractor piggy-backed. It makes an excellent compliment to the 10-inch LX200.

as the camera will not swing through the fork mount and hit the telescope. This can damage the Dec drive and/or the camera. If used in Alt/Az mode you would have the same problem, near the sky zenith. Overall, it's a great telescope. Recently, the telescope had developed a communication problem. Its electronics and motors will need to be disassembled and sent back to Meade for repairs. When dealing with state-of-the-art components, problems can almost be expected to arise.

The Camera

The Meade Pictor 416 CCD camera is well matched for the 10-inch LX200; but for deep-sky objects I suggest using a focal reducer. The Optec Maxfield 0.33 reducer makes the combo of telescope and CCD camera excel. The camera has been upgraded to the 416 XTE, which has the new enhanced blue sensitive KAF-0401 E Kodak chip. This is a worthwhile upgrade, and like the new Pictor XTEs, it contain the new chip. The camera has worked quite well over the past four years. It has had to be repaired twice, once for a faulty shutter, the other time for dew build-up on the CCD chip. CCD chips operate at extremely cold temperatures. If moisture gets inside the camera, it will form dew on the chip.

The Computer and Software

The computer you purchase should match the software you wish to use. I had tried to run a computer outside, in the dome, but cold temperature seriously affects the performance of the computer. Even using a laptop, it is a struggle to keep it warm so that it would run properly. It is much more convenient to operate the observatory from the comfort of my home. The camera software has improved dramatically, and so has the telescope control software. I like to use The Sky IV by Software Bisque. There is also a variety of great software packages for image processing. I would suggest trying many demo packages to see what suits your needs.

Galaxy mag 22.3

Figure 10.12. In only 24 minutes this set-up can achieve 22nd magnitude.

Capabilities

In 1998 the Arcturus Observatory participated in The Deep Field Challenge, presented by *Sky & Telescope* Magazine in its May 1998 issue on page 119. It was part of an article by Bradley E. Schaefer entitled *Limiting Magnitudes for CCDs*. The purpose of the "challenge" was to find how faint a magnitude amateur astronomers could image. Our goal was not to win, as our telescope most likely would not have the largest aperture, but to determine our observatory's capabilities within a reasonable time allotment for an image. With a CCD image of 24 minutes duration, we were able to achieve a magnitude of 22.3. So with a 10-inch telescope and a CCD camera, this observatory was able to exceed Palomar's Sky Survey, which was done with film. This is why we can consider this a large observatory with a very small footprint.

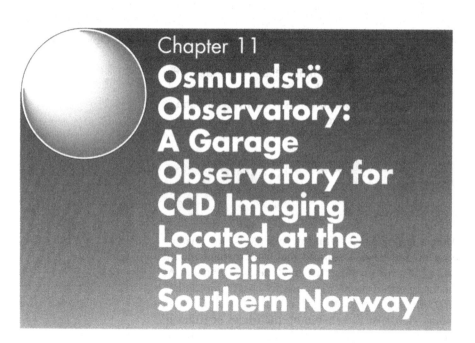

Chapter 11

Osmundstö Observatory: A Garage Observatory for CCD Imaging Located at the Shoreline of Southern Norway

Alf Jacob Nilsen

"White snow, twinkling stars on a crystal clear black sky and the sound of fishing boats leaving for the sea", that is a memory of my childhood. I could stand for hours watching the stars and wonder what is was like out there. My first telescope was a 60 mm spotting scope. Later came a 2.4-inch Unitron refractor, which is still in use. In my teenage years my parents bought me a 6-inch Charles Frank equatorially mounted Newtonian reflector, and I made my first attempts at astronomical photography. At that time, back in the late 1960s, no CCD cameras existed. The central stars of planetary nebulas and remote galaxies were all just a dream. Only the professional guys could obtain pictures of such objects. I kept *Sky & Telescope* next to my bed, read the articles about amateur astronomical photography, and never stopped dreaming about one day being able to shoot such stunning deep-sky photos myself.

Twenty-five years passed, I went away for studies and work, married and had a family, and there was little time for astronomy. But then came the CCD-revolution, the children grew up and things changed. Suddenly the deep sky was within the reach of an amateur's equipment. I was hooked on astronomy again, and it is more excited than ever.

The Site and the Possibilities

After a few sessions setting up my LX200 for imaging on a nightly basis, I realized that a permanent mount was a must for imaging with my type of equipment. As a consequence I began planning my observatory in late 1997. My site at 58° 14' 5" N, 6° 32' 12" E is the small fishing village of Kirkehamn on the island of Hidra in the south of Norway. The location is somewhat unusual for hosting an astronomical observatory. Located not more than 100 metres from the North Sea, and at an elevation of about 3 metres (10 feet), it is certainly not the very best place in the world for CCD imaging (Figure 11.1). The location is often both humid and windy, and there are relatively few clear nights each year. However, when the sky decides to go clear, the seeing can be really good and even extraordinarily good, but such night are limited to less than twenty in a year. As if this were not enough, the horizon is also limited, and there is some light pollution from street lights.

Could a CCD camera function in such a place? Would it be worth the effort and money to build an observatory here? What options did I have? Norway is full of mountains where light pollution is still almost non-existent, and it is not difficult to find a remote, dark spot within an hour's drive from where I live. However, in order to learn CCD techniques, I felt it would still be better to have an observatory within reach close to my home than to travel away on clear nights. A permanent building away from home was never an option, so a remote site would have had to be on a nightly set-up basis, which I had tried unsuccessfully many times before. A permanent observatory in my garage next to the house became the final choice. A portable second set of imaging equipment that could occasionally follow me to the mountains became a wish for the future.

Figure 11.1.
Osmundstö Observatory is located just a hundred yards from the shoreline of the North Sea in the small fishing village of Kirkehamn on the island of Hidra in south-western Norway.

The Construction

Anyone who has visited the coast of south-western Norway knows how fast the weather can change. In a matter of minutes a rain shower can burst out of the "clear" sky. Therefore an observatory located here must be easily operated and allowed to be open and closed in minutes. It also had to be wind- and waterproof and insulated to allow a nearly constant in-door temperature during the whole year. My equipment was a 10-inch Meade LX200 Schmidt–Cassegrain mounted on the Meade Superwedge. As I saw the possibility of changing to a slightly larger and better-quality scope later on (such as a 12.5-inch Ritchey Chretien), the observatory should be big enough to support such an instrument. Finally it should be constructed so that it would give an optimum view when fully opened.

The roof of the garage had an angle of about 22.5° (Figure 11.2). In order to get the maximum view to the

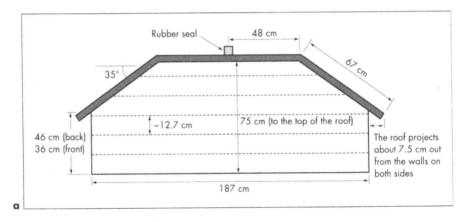

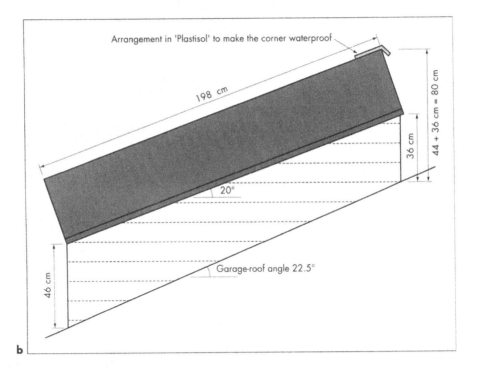

Figure 11.2. The part of the observatory that projects above the garage rood, **a** As seen from the lower end. Plastisol plates and styropore make up the 5 cm (2 inches) thick, insulated roof-halves. The angle of the observatory roof is about 35° and the roof projects about 7.5 cm (3 inches) out from the walls on both sides. The dashed lines represent 5-inch (14.7 cm including 2 cm overlap) wooden waterproof planks, a type much used in western and southwestern Norway. Internal opening: 169 × 169 cm (5 feet 8 inches × 5 feet 8 inches); external opening: approximately 180 × 180 cm (6 × 6 feet). **b** The observatory asseen from the western side.

south and at the same time use the garage roof as a shield from a couple of streets light, the observatory had to be raised about 80 cm (2 feet 8 inches) above the garage roof. The observatory's roof would have an angle of 20°, which was slightly less than that of the roof of the garage.

A rotating dome and a sliding roof were both considered, but after listing all the pro and cons I decided that two trapdoors folding above each other would be the best solution to the problem of opening the roof easily. The right door would close over the left one with rubber packing in between, making the doors waterproof (as outlined in Figures 11.3–11.5). In order to maintain the maximum view from inside the observatory, the doors have to close just an inch or so above the top of the telescope.

I like to do carpentry work myself, and I had no worries about the interior of the building. Insulating

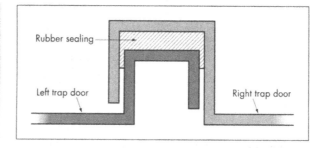

Figure 11.3. The rubber seal that closes and seals the trapdoors.

Rubber sealing

Left trap door

Right trap door

Figure 11.4. The right trapdoor folds over the left one and is sealed where the finger is pointing.

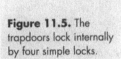

Figure 11.5. The trapdoors lock internally by four simple locks.

the building, setting up the internal walls, building the pillar and mounting the optical equipment would be "a piece of cake". Cutting the garage roof, setting up the outside observatory walls and building and mounting the trapdoors was, however, another story. Two good friends, my long-time mate, neighbour and very skilful carpenter Johnny, did the carpentry work, while another friend and local tinsmith, Werner, designed and made the trapdoors. Without their help the observatory would never have been a reality. Many thanks to both of them! (Figures 11.6–11.9)

The trapdoors had to be lightweight, but at the same time strong and insulated. The doors measure (48 + 67) × 198 cm (3 feet 10 inches × 6 feet 7 inches) each (Figure 11.2), which makes them roughly 1.3 m² (13.5 square feet) in size. They are made of a material known as "plastisol" in Norway. Plastisol consists of thin (normally 1 mm) steel plates coated with plastic of

Figure 11.6. The observatory with closed trapdoors seen from the top of the garage roof.

Figure 11.7. The contruction began by cutting the original roof in the garage.

Figure 11.8. The outside walls were raised above the roof of the carport. Black plastisol plates are mounted to form a drain for rainwater on three sides of the projecting walls.

Figure 11.9. The author himself did the internal carpentry work, and the observatory gradually took shape.

selected colours. We used black plates on the outside and white on the inside. The doors were made 5 cm (2 inches) thick and contained polystyrene for insulation. Four thin steel beams (profiles) run transverse inside each door to increase the strength. Each door projects 7.5 cm (3 inches) beyond the walls and is fastened with four steel hinges (Figures 11.10 and 11.11). This created a 1 cm (0.4 inch) opening under the doors, between the trapdoors and the sidewalls of the observatory. I planned to seal this opening with a rubber seal, but this has proved to be unnecessary. The top front corner is a critical point and a difficult spot to seal. Here a bracket (Figure 11.12) is mounted to prevent raindrops from entering the observatory this way. The roof is absolutely tight and not a single drop of water or a single snowflake has entered even during the worst storms (which I can assure you are pretty ugly!). The doors are furthermore so light that one person can easily open and close them in a matter of seconds.

The walls projecting above the garage roof are 5 cm (2 inches) thick and are covered on the outside with wooden waterproof planks of a type much used in western and south-western Norway and locally named "Vestlandskledning". These, as well as the rest of the observatory walls, are insulated with tare board, rockwool and thin plastic. An arrangement of plastisol was made outside and around the walls

Figure 11.10. Four steel hinges mount each trapdoor.

projecting above the garage roof to allow rainwater to be drained away from the observatory and into the gutter (Figure 11.8). This has worked perfectly.

Figure 11.11. The trapdoors project about 7.5 cm (3 inches) outside the walls. Note the placement of one of the hinges when the door is closed.

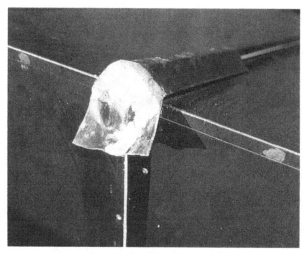

Figure 11.12. This protective cap, made from plastisol, protects from leakage in the difficult upper corner of the roof where the two trapdoors meet.

Interiorly the walls are covered with 5-inch (12.5 cm) pinewood. The floor was originally made of concrete, but a second floor made from 10 cm (4 inch) polystyrene and 2.5 cm (1 inch) thick wooden plates was laid down, floating above the original floor. The wooden floor was built independently of the telescope pillar and separated from the pillar with foam to minimize vibrations.

The pillar measures 120 × 40 × 40 cm (48 × 16 inches × 16 inches) and is made of concrete and anchored to the rocky ground below the concrete floor with 20 mm (0.8 inch) and 12 mm ($\frac{1}{2}$ inch) iron bars (Figure 11.13). Although vibrations have not been a problem so far, it would definitely have been better to have anchored the pillar deeper in the ground by digging a deep hole filled with iron and concrete. This was, however, not possible and was never considered to be realistic in this project. An outside-diameter 18 × 80 cm (7.2 × 32 inch) steel tube runs down in the pillar and projects 45 cm (18 inches) above the wooden shelf. An inside-diameter 18 × 35 cm (7.2 × 14 inch) steel tube is locked with three 8 mm ($\frac{3}{8}$ inch) setscrews outside on the main steel tube and has a baseplate on its top that fits the Meade Superwedge (Figures 11.14–11.16). The space between the two tubes is filled with frost-resistant, high-quality grease lubricate. The complete arrangement allows a rough polar alignment as the outer tube can be rotated about 60° in each direction. 220 V AC is available in a narrow cupboard on one side of the pillar, a cupboard that also houses the power supplies.

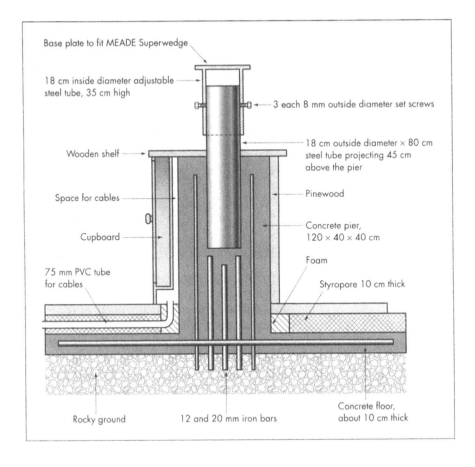

Base plate to fit MEADE Superwedge

18 cm inside diameter adjustable steel tube, 35 cm high

3 each 8 mm outside diameter set screws

18 cm outside diameter × 80 cm steel tube projecting 45 cm above the pier

Wooden shelf

Space for cables

Pinewood

Cupboard

Concrete pier, 120 × 40 × 40 cm

Foam

75 mm PVC tube for cables

Styropore 10 cm thick

Concrete floor, about 10 cm thick

Rocky ground   12 and 20 mm iron bars

A 60 cm (2 foot) bench runs along the entire western wall of the observatory and gives room for the computer screen, maps, etc. (Figure 11.17). Below the bench are drawers and cupboards for housing optical and technical equipment as well as a computer. All cables and connections from the computer to the telescope and CCD camera are drawn through a 75 mm (3 inch) PVC pipe running below the floor. The cables run behind the pillar-cupboard and through a small opening in the wooden shelf to the telescope and camera. Osmundstö Observatory measures 280 × 215 cm (9 feet 4 inches × 7 feet 2 inches) giving it an area of 6 m^2 (65 square feet). The opening towards the Universe is 2.7 m^2 (162 × 165 cm). I just love it when my scopes point towards the deep sky and I was a very lucky and happy man when the observatory saw its "first light" on 28 November 1998.

Figure 11.13. Shape and measurements of the pillar.

Figure 11.14. the pillar consists of an 18 × 45 cm (7.2 × 18 inch) outside diameter steel tube mounted in a concrete socket anchored to the concrete floor. Another 18 cm (7.2 inches) insides diameter steel tube is threaded outside the anchored tube and locked by three setscrews after a rough polar alignment is done. This photograph shows the mounting of the tube, which is made perfectly level.

The Optical Equipment and Modifications Done

My telescopes are a Meade LX200 10-inch, f/6.3 Schmidt–Cassegrain with a 4-inch, f/10 Fujigawa achromatic refractor mounted piggyback (Figures 11.18–11.20). When I bought the LX200 back in 1995, I intended to use the equipment for conventionally astrophotography and visual observing. Gradually I turned towards CCD-imaging, and had I known then what I know now, I would perhaps have chosen differently. As an all-round scope and as a scope for learning imaging techniques, as well as for combining imaging with visual observations, the LX200 is a great tool! However, the general optical quality, mirror shift and the relatively poor design of the drives put limits on what can be achieved with respect to long time

Figure 11.15. The Superwedge is mounted on the outer steel tube for testing.

exposures. Although an LX200 can – if collimated, calibrated and adjusted carefully – produce high-quality images, it will never perform to give results like those than can be achieved with a Ritchey Chretien or an Astro Physics apochromatic refractor (but of course the price of equipment must also be taken into consideration).

In order to optimize the scope for my purposes, I have modified and added some equipment to the original set-up. The balance system, as shown in Figure 11.21, was adjusted by mounting 10 mm (0.4 inch) screws and counterweights which can be adjusted to carefully balance the scope. A simple mirror lock (originally constructed by Chris Vedeler) consisting of a 0.25-inch (6 mm), 6-inch (150 mm) long screw, a 0.2-inch (5 mm) nut and a 3-inch (75 mm) long spring mounted in the transport-screw fitting at the rear end of the scope, help prevent mirror shift during an image session. A home-made clock-scale added to the back-

Figure 11.16. The telescopes are mounted and the trapdoors are closed.

plate of the scope (Figure 11.22) helps orientate the camera to a position where a relatively bright guide-star can be found. The original focus control on the LX200 was equipped with a focus counter from Jim's Mobile Inc., which is handy for logging a rough focus position. A van Slyke Micro Gage MGF2 manual focuser secures accurate and fine-tuned focus, which is a must for CCD imaging (Figure 11.23). This piece of equipment is absolutely top quality and so is the optional van Slyke Versa Slider port, which gives me the possibility of viewing the target prior to imaging without altering the camera. A van Slyke 2-inch filter slot can also be mounted as part of the equipment configuration. However, please remember that a lot of accessories added to the scope make it too heavy, increases the focal length and narrows the field of view (FOV) greatly.

As this book goes to press I have also started experimenting with the Temperature Compensating Focuser (TCF) from Optec Inc. This device automatically compensates for the shift in focus as the temperature changes during an image session. By the use of various imaging software (such as MaxIm ver. 3.0 or CCDSoft) perfect focus – both in luminance as well as in RGB imaging – can be obtained from the PC, a

Figure 11.17. Internal decoration consists of a bench, shelves and cupboards with drawers to house optical and technical parts as well as a computer.

feature that makes the focus procedure much simpler than time-consuming manual focus adjustments. As "perfect focus" is one of the most critical parameters in imaging, it is my hope that the Optec TCF focuser will improve the results even more.

Collimating the scope almost perfectly has a great impact on the image quality. I have tested laser collimators, but find them hard to use for fine-tuning collimation. Personally I prefer the "old-fashioned" star-collimation method. To make the adjustments easier I have replaced the three original collimation screws with bigger and more easily adjustable screws made by Chris Heapy, UK. I advise every LX200 owner to do this as it makes the adjustments so much easier. Dew is sometimes a big problem here by the ocean, and on such nights the Kendrick Dew Remover System helps a lot. However, heat from the system can disturb the image quality and I therefore use the system only on nights when dew is really a big problem.

Figure 11.18. The telescopes with the CCD equipment point towards the sky.

The Superwedge has also been modified slightly by replacing the original steel wrasses with Teflon wrasses, which smoothes out the forces a bit and makes adjustments easier. The original Meade Superwedge is not easily polar-aligned perfectly, but for a permanently mounted site it can gradually be adjusted to an accurate alignment. Once the level and the polar alignment are perfect, leave the wedge alone! I have used both Star Drift and the T-point software made by Software Bisque to check the polar alignment, and have obtained an alignment that shows no star-drift for at least 20 minutes observed at high magnification through a 9 mm reticule eyepiece. I have also marked the adjustment wheels on the wedge in order to be sure that the wedge is adjusted the correct way when the star-drift method is performed. Simple "west/east" and "raise/lower" labels are really helpful.

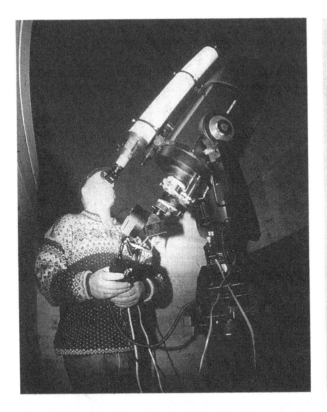

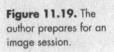

Figure 11.19. The author prepares for an image session.

The Imaging Equipment, Hardware and Software

In the early 1970s, when I made my first attempts at astronomical photography, the scope was guided manually – a freezing and very boring process. When Dennis di Cicco reviewed the SBIG's autoguider and simple ST4 imager in the September 1990 issue of *Sky & Telescope*, the hope of more relaxing guiding arose. This article was my first introduction to any CCD device, which resulted in the purchase of the ST4 from SBIG in December 1990. The device was to be used as an autoguider for conventional deep-sky astrophotography before the scopes were permanently mounted. The ST4 is a great piece of equipment and really set the SBIG standard back in the "early days" of imaging and autoguiding. Although I do not use the ST4 much these days, I am never going to get rid of it. For a short period, also before the scopes were permanently mounted, I used the ST-6 CCD camera, but soon sold

Figure 11.20. The telescopes seen from the outside.

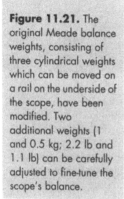

Figure 11.21. The original Meade balance weights, consisting of three cylindrical weights which can be moved on a rail on the underside of the scope, have been modified. Two additional weights (1 and 0.5 kg; 2.2 lb and 1.1 lb) can be carefully adjusted to fine-tune the scope's balance.

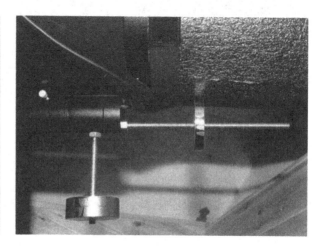

Figure 11.22. A simple "clock-scale" is glued on the rear end of the scope and is useful for orienting the CCD camera.

Figure 11.23. One of several configurations used with the LX200. From right to left: The Meade LX200 (f/6.3), van Slyke focuser MGF2, van Slyke Versa Slider, van Slyke filter slot, SBIG AO7 and ST7E. This set-up increases the focal length from the original 1600 to 2100 mm, which gives the scope an f-stop of 8.3 and a field of view on the ST7E KAF400 chip of 11′ 18″ × 7′ 32″. This is a good set-up for imaging tiny galaxies where visual observations can be done through the parafocal view in the Versa Slider port. Alternatively the filter slot can be placed in front of the slider port.

it and bought the ST7 model with self-guiding options. In the summer of 2000 the camera was upgraded to the more light-sensitive ST7E ABG version. Coupled to the ST7E is the SBIG AO-7 Adaptive Optics, a piece of equipment that allows autoguiding corrections to be performed up to 100 times a second, correcting for atmospheric turbulence as well as for periodic errors in the drives. Since its arrival on the scene, discussions about its pros and cons and what it can do and cannot do have flourished on the Internet and in various magazines. I am not even going to try to outline the details here, but only state that to me the device is almost magic. The use of the AO-7 Adaptive Optics from SBIG has really improved my results a lot and I use it for almost every session. Also available is the SBIG CFW-8 colour-filter wheel, which allows the possibility of colour imaging through RGB filters. An STV from SBIG was installed in December 2000 and acts as an electronic finder by the use of the FR-237 focal reducer that comes with the system. This unit is also planned to work as an autoguider with my up-coming portable equipment based around a Losmany G-11 equatorial mount. The STV is also great for imaging the Moon (Figure 11.24) and planets through its "best sharp" function, when the unit automatically takes a series of images and stores only those that are

Figure 11.24a. A mosaic of two images of the Moon taken on 1 May 1 at 23.13 local time using the "best sharp function" in the STV and the Fujigawa 4-inch, f/5.95 achromatic refractor.

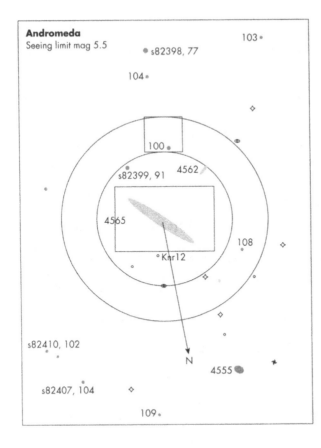

Andromeda
Seeing limit mag 5.5

103
s82398, 77
104
100
s82399, 91 4562
4565
108
Knr12
s82410, 102
N
4555
s82407, 104
109

Figure 11.24b. The ST7 is a self-guided CCD camera housing both an image chip (KAF400) and a guiding chip (TC211). In order to obtain good tracking a bright star (preferably brighter than magnitude 10) must fall inside the guide-chip's FOV. If the FOV-indicator is displayed in *The Sky* software, potential guide-stars are easily located to the guide-chip's FOV (small square) and by rotating the N-indicator the orientation of the camera can be found. I have found this too to be most valuable for planning image sessions and I have made an archive of FOVs for my most-used configurations. Image from *The Sky*, by courtesy of Software Bisque, Inc.

sharper than the previously stored one. I also use the STV for measuring the evening's seeing by using the "Differential Image Motion Monitor" (DIMM) method, which is also used by serious amateurs and gives an accurate log of seeing through the 10-inch LX200 SC. So far my best seeing has been 1.1 arc seconds measured on 24 March 2001. The average seeing after measuring 31 nights is 2.25 arc seconds.

It is of course not possible to use all the available equipment at the same time. The equipment does give me the possibilities of constructing different configurations for different objects and needs (Figure 11.25). The focal length and field of view are altered when the configuration is changed. Table 11.1 gives an overview of the most used set-ups and their focal lengths and fields of view.

I use a simple Pentium 166 MHz computer for operating the CCD camera and the necessary software used. The computer is easily powerful enough for this purpose. The CCD camera is operated with the MaxIm

Figure 11.25. Another configuration, from right to left: The Meade LX200, f/6.3, van Slyke focuser MGF2, van Slyke filter slot, SBIG colour filter wheel CFW-8, AO7 and ST7E. This set-up increases the focal length from the original 1600 to 1993 mm, which gives the scope an f-stop of 7.9 and a field of view on the ST7E KAF400 chip of 11′ 55″ × 7′ 56″.

DL/CCD software from Diffraction Limited, a program that also has, among other great camera control options, the very useful auto-focus features. *The Sky* level V from Software Bisque is used on every session. The LX200 corresponds easily with the program and I normally operate the scope via the computer. With accurate polar alignment, I have little problem letting the LX200 slew relatively long distances over the sky and still place the selected object well inside the FOV of the image chip. The software allows a FOV-indicator to be made, matching the actual configuration used, and to display it on the screen. In this way the exact FOV for both the imaging chip as well as the autoguider chip in the ST7E are shown on the virtual sky. By displaying the "rotate tool" (yet another of *The Sky*'s neat features), one can easily determine how the camera must be rotated in order to locate a suitable guide-star. All this can be done during cloudy evenings. By saving each file one can make an archive of potential objects for imaging, their FOVs, camera orientation and guide-star location (Figure 11.24). Rotating the camera relative to the home-made clock-scale at the rear end of the scope, according to the position displayed on *The Sky*, does normally easily locate the guide-star on the guiding chip.

Furthermore, I have found it important to take dark frames for every image session, and usually take a set of 8–16 dark frames previous to an evening. Flat fields are taken immediately after imaging an object by the use of

Table 11.1. Field of view (FOV), focal length (FL) and f-stops for some commonly used equipment configurations

Configurations			
LX200 with van Slyke MGF2 focuser and	**FOV** (arc minutes)	**FL** (mm)	**f**
Focal reducer f/6.3 – A07 – ST7E	21.3 × 14.2	1109	4.4
Versa Slider – A07 – ST7E	11.6 × 7.6	1993	7.9
CFW8 – Focal reducer f/6.3 – A07 – ST7E	19.2 × 12.6	1226	4.9
Versa Slider – CFW8 – A07 – ST7E	11.1 × 7.3	2099	8.3
LX200 with Optex TCF and			
CFW8 – Focal reducer f/6.3 – A07 – ST7E	20.0 × 13.2	1189	4.7

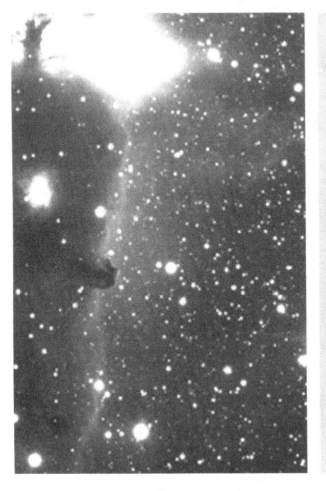

Figure 11.26. The small 60 mm apochromatic refractor FS60-C from Takahashi has recently been added to the observatory's equipment. It is great for wide-field imaging as shown here where the dark Horsehead Nebula is captured against the emission nebula IC434 in Orion's sword. The bright star at the top is the 1.7 magnitude "Alnitak". The image from February 2002 is a composition of 8×600 second frames with the small refractor mounted piggyback on the LX200.

a "flat-field box" (Figure 11.27) which is mounted in front of the telescope. It is necessary to take a series of flat-frames, and I usually take 8–20 frames for each colour. Series of "flat darks" are also taken (frames taken into the flat-field box, with the illumination switched off). *The Handbook of Astronomical Image Processing* (by Richard Berry and James Burnell, Willmann-Bell Inc., 2000) gives excellent descriptions of how to build a flat-field box and how to shoot dark and flat frames, and includes a good package of image processing software.

During 2001 I have experimented with very long exposures, such as more than 30 integrations of M51 (Figure 11.28), each lasting for 10 minutes, giving a total exposure time of more than 5 hours. By this method, I have been able to reach very deep into the Universe and have recorded stars of magnitude 20 and

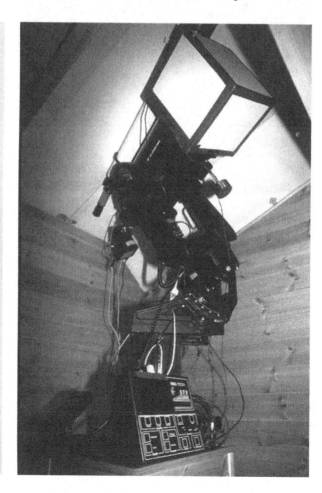

Figure 11.27. The home-made flat-field box mounted in front of the telescope. The box is made from cardboard and illuminated with four 2.5 V DC lamps. A DC converter and regulator installed in the PS cupboard makes it easy to adjust the illumination inside the box.

Figure 11.28. M51 and NGC5195 in Canes Venatici. A 5 hour 10 minute image taken with the ST7E through the 10-inch, f/4.8 LX200 on 22–23 April 2001. The following record shows the faintest star in the image (R-magnitude 20.72) recorded by the use of the astrometry and photometric functions in the Astroart software (by Mr. Arne Danielsen, Oslo). The mag. value is a little uncertain as the signal-to-noise ratio is very weak.

Xc	Yc	Ra°	Dec°	Mag	E_mag	FwhmX	FwhmY	S/N	Ra	Dec
411.91	140.08	202.590140	47.211505	20.72	0.41	1.08	1.06	6.40	13 30 21.634	+ 47 12 41.42

A total of 292 stars were recorded in the image and at least 50 weak galaxies are also detectable in the field, which covers about 13 × 20 arc minutes.

fainter. By the same method, more than 235 galaxies of the Coma Galaxy Cluster were recorded (Figure 11.29) in a 4 hour 10 minute exposure. When I began taking

Figure 11.29. The Coma Galaxy Cluster (Abell 1656) in Coma Berenices. A total of about 235 galaxies were counted in the field, which covers 00 19′ 58″ × 00 13′ 06″. The image centre is at RA 12h 59m 46.4s, Dec +27 55′ 02″ (epoch 2000). Scale: 1.58 arc-seconds/pixel, north angle: 2.14. A 4 hour 10 minute (25 × 600 seconds) exposure with the ST7E through the 10-inch, f/4.8 LX200 SC on 20–21 April 2001.The image was processed in AIP$_4$WIN with: Fast R-L Deconv, Blur = Gaussian, Rad = 1.0, NR = 0.5, Iter. = 5. Only high-frequency components were processed during deconvolution. Gamma-log scaled, Min = 7000, Max = 65 534, Gamma-log = 0.225, Gaussian histogram shaped, sigma spread = 4, Peak skew = 0.15.

CCD images, I usually wanted to take as many objects as possible on a night. After a few years of experiences I have found it much more useful to concentrate on one or two objects a night and to try to obtain really high-quality images.

So far I have shot mostly images of deep-sky objects. After a night of imaging I usually end up with a series of images of one or maybe two objects and a lot of calibration frames, all of which have to be processed later. In processing the deep-sky images I use software packages such as CCDSHARP, MaxIm DL/CCD, MIRA AP, AIP$_4$WIN, ASTROART and Adobe Photoshop. One of the neat things with imaging is that you have the option of playing with your images on cloudy nights (of which there are plenty here).

Typical Procedure for an Image Session from Osmundstö Observatory

I have found that careful preparation and planning ahead of the deep-sky image session as well as a fixed procedure during the session are important factors for success. My procedure usually consists of:

1. Deciding which deep-sky object to shoot tonight, selecting the configuration and mounting the necessary equipment on the telescope.

2. Activating the camera, setting the temperature (normally −15 °C or −25 °C) and taking 8–16 dark frames before it gets dark.

3. Checking tonight's seeing using the STV and its DIMM function.

4. Checking the necessary camera position according to the FOV displayed in *The Sky* and rotating the camera to the actual position using the clock-scale on the scope.

5. Checking the balance and fine-tuning the "backlash correction".

6. Doing a rough focus using a Hartmann mask on a bright (magnitude 3) and medium bright (magnitude 6–7) star.

7. Fine tuning the focus using a relatively weak star and the *focus/inspect* options in MaxIm DL/CCD. Making notes on the focus positions or, using the TCF, run an auto-focus routine.

8. Centring the object on the main chip.

9. Checking that the guide-star is inside the guide-chip's FOV as planned.

10. Calibrating the ST7E and AO7 on this position.

11. Recentring the object on the image chip if necessary.

12. Taking a 60 or 120 second test exposure to check that everything tracks well, that the object is centred as planned and that the focus is OK.

13. Starting the luminance exposures (normally a series of 600 seconds each).

14. Taking the L-flat fields using the flat-field box.

15. Selecting the colour filter and refocusing for RGB images individually by adjusting the MGF2 micro-

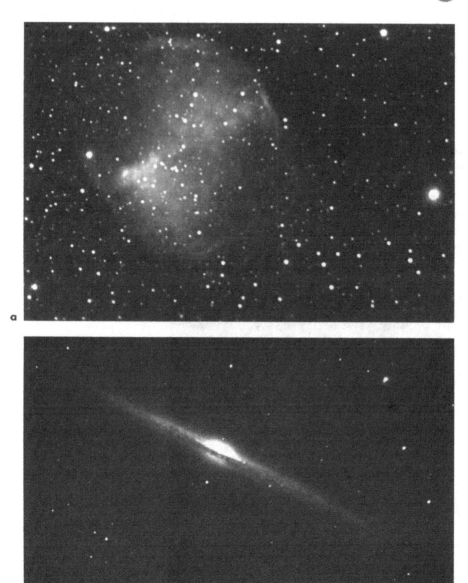

Figure 11.30. Four images from Osmundstö Observatory: **a** 1800 second image of M27 in Vulpecula; **b** 1200 second image of the edge-on galaxy NGC4565 in Coma Berenices; **c** 1200 second image of the Crab Nebula (M1) in Taurus; **d** The Siamese Twins, two colliding galaxies (NGC4567 and NGC4568) in Virgo, here in a combination of a 900 second and a 1200 second image. *(Figure 11.30 c and d on the following page)*

meter scale or refocusing using the TCF and an auto-focus routine.

16. Starting the RGB images (normally a series of 600 seconds each).

17. Shooting a series of RGB flats after each colour session.

Fine-tuning the focus is crucial. Getting this right, in combination with a night of good seeing, is crucial for obtaining "the ultimate image". Obtaining good flat fields is for me by far the toughest task. In order to obtain good flats you need to have focused carefully and to have the camera in exactly the same position as used when the image is shot. This can seldom be achieved during twilight time and therefore the twilight

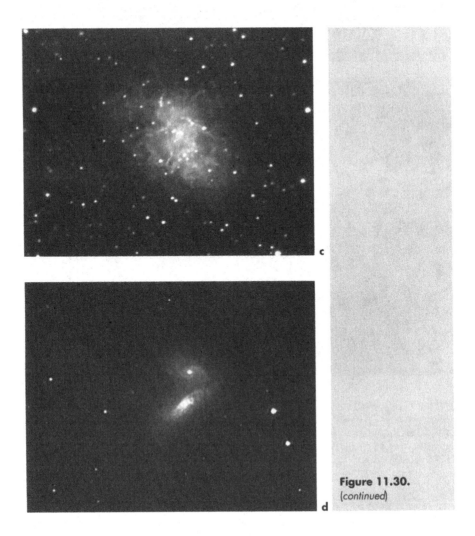

c

d

Figure 11.30.
(continued)

sky cannot be used to obtain flats (unless the camera is left with the same focus and orientation till the next clear night). Therefore a flat-field box is for me the best way to obtain flat frames. Refocusing when inserting the filter is absolutely essential. The van Slyke MGF2 focuser does, however, have a micrometer screw with a microscale that enables the refocus to be estimated and done in a short time. I make notes on how the focus is obtained, as well as logging all the image parameters. Such information is most useful over time.

Conclusions

Osmundstö Observatory has so far been a success. Since the construction was completed it has been used on almost every clear night. It has made deep-sky imaging possible even on the edge of the North Sea. Would I have done anything differently? Perhaps I would have raised the building more in order to get a better horizon towards the south. On the whole I am, however, very satisfied! The "deep images" obtained with several hours of exposure, done in Spring 2001, have taught me a new way of using my astronomical imaging equipment, and have given me experience that I shall surely develop further in the future. I plan to gradually change the optical equipment to become even better and to get a second set of portable equipment that can follow me to the mountains and to southern countries, but these are all goals well ahead into the future.

While the image is being taken and the scope is tracking perfectly, I usually leave the site, to avoid disturbing the equipment and cause vibrations. The equipment seems to do best alone! If conditions are excellent and no rain can be feared, I even go to sleep and let the technique work peacefully alone.

A cup of coffee, maybe a glimpse at the late-night news, a short nap and logging the night's seeing, are useful tasks to do while the CCD opens the deep-sky for you and brings you closer to a Universe that was only a dream a decade ago!

For more information on Osmundstö Observatory please take a look at http://hjem.sol.no/alfnil/

Chapter 12

Huntington Observatory, York

Mike Brown

How it Came About

I built my first telescope, a 10 cm, f/10 Newtonian in 1963 and upon completion quickly became aperture hungry, passing rapidly through 23 cm, 30 cm and finishing up in early 1966 with a 37 cm, f/5.75 Newtonian. The telescope was on a motorized equatorial mounting to enable me to follow my rapidly developing taste for photographing the Moon and planets. What I hadn't realized when I embarked on the building of this instrument was just how heavy and unmanageable it would be and the movement of it in and out of my garage became a major physical problem. In addition the lining up of it each time on the approximate pole also became a real chore.

By mid-1966 I'd had enough and decided that a permanent observatory was the only answer. It would avoid the use of muscle power, eliminate the lengthy polar alignment, and give me quick availability with 100% protection for the telescope. In addition there would be a lessening effect from wind vibration and it would provide an improved personal environment during the cold winter months. The site I earmarked in my back garden would allow unobstructed views of the sky from east through to west apart from an eight-degree loss on the horizon due to my own and surrounding properties. This latter item was of no consequence as with the seeing experienced here in York I rarely observed anything below thirty degrees or so.

Figure 12.1.
Observatory from the
west side.

Preliminaries

Before starting construction I thought it wise to approach my local authority to establish if planning permission was necessary and if building regulations had to be met. I envisaged a permanent structure with proper foundations, a breeze block wall topped with a rotating galvanized steel dome. The response was quick, planning permission was not required but building regulations had to be met and would I please supply them with two copies of a drawing of the proposed structure.

This I did and almost by return was told that what I proposed met the regulations and work could commence but it would have to be inspected when the walls were erected and finally upon completion of the observatory.

Construction

I started off by digging out for the foundations a circle of inside diameter 3.5 × 3.9 m (11 feet 4 inches × 12 feet 7 inches) outside diameter by 0.4 m (16 inches) deep. I calculated that this was just about right for a structure with 3.7 m (12 feet) diameter walls, five breeze

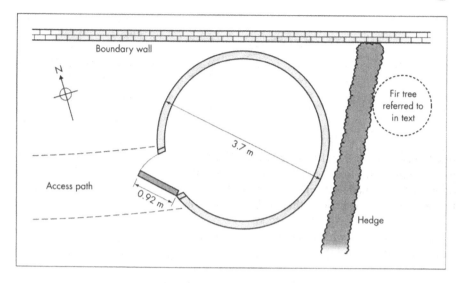

Boundary wall

N

Fir tree
referred to
in text

3.7 m

Access path

0.92 m

Hedge

Figure 12.2. Plan view
of the observatory
located at 53 59′ 12″
N, 1 03′ 20″ W.

blocks high, i.e. 1.3 m (4 foot 3 inches) tall supporting a
galvanized steel dome of slightly greater diameter and
1.9 m (6 feet 2 inches) to the highest point.

One cubic metre of foundation-grade ready mixed
concrete was delivered and I had the arduous job of
wheel-barrowing it from the end of my bungalow drive
to the site in the back garden, a distance of some 60 m
(190 feet). Using previously levelled wooden dowels I
managed to accurately lay the foundations. Sufficient
concrete was delivered to enable me to make an access
path to the site from an existing path in the garden.

Now for the walls: breeze blocks 46 × 23 × 7.7 cm
(18 × 9 × 3 inches) were laid five rows high.
Twenty-five blocks would be required to make the
approximately circular shape but I omitted two blocks
per row to make provision for a doorway on the west
side into the building. The walls and doorway make a
twenty-four sided structure with the door space being
92 cm (37 inches) wide. To ensure that the blocks
formed the closest possible representation of a circle I
placed a 2.5 cm (1 inch) steel tube in the centre of the
site and fastened a length of steel wire about 1.77 m
(5 foot 9 inches) long which looped over the tube and
would therefore rotate, enabling me to get the centre of
each block exactly the same distance from the centre of
the site. The blocks were cemented in place and a
careful check was kept to ensure levelness and vertical
integrity were maintained.

After the five rows of 23 blocks were erected a
wooden frame for hanging the entrance door from was

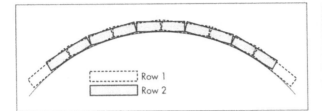

fitted, made out of 5 × 10 cm (2 × 4 inch) timber, the top section being level with the top of the wall. The door itself was made out of 1 × 2 cm (0.4 × 0.8 inch) timber for the frame and then skinned with 8 mm ($\frac{3}{8}$ inch) thick exterior-grade plywood.

For the rail on which the dome would rotate I had thought of using 2.5 × 2.5 cm × 3 mm (1 × 1 × $\frac{1}{8}$ inch) thick T-section steel bent to form a circle of the appropriate diameter so it could be fastened to the centre of the breeze block wall. Enquiries made locally didn't come up with this material so I had to use 25 × 3 mm (1 × $\frac{1}{8}$ inch) thick hot rolled steel flat material to be fastened to the wall top with cast aluminium T-section brackets. Fortuitously I had already taken steps to produce a much more rigid German mounting for the 37 cm and

Figure 12.4. Part side view showing the half-block offset for each row of breeze blocks.

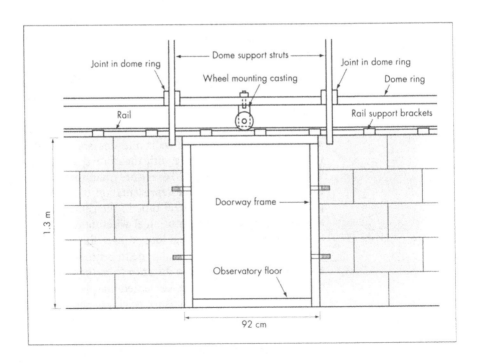

intended to produce my own castings for it. To do this I had therefore built a gas-fired crucible furnace, which I could now use for casting from wooden patterns for the aluminium T-section brackets, the subsequent machining of them being undertaken on my lathe and vertical milling machine in the workshop.

Some 75 brackets were required and I cast them in strips of three, which were subsequently separated and slotted to accept the 3×25 mm ($\frac{1}{8} \times 1$ inch) flat steel strip. The centre of each bracket was kept equidistant from the centre of the building using the pole and wire technique adopted for centralizing the walls. The brackets were Rawlplugged to the top of the wall. When all were fitted I found that the lengths of steel strip slotted in them automatically made a circle of the correct diameter without any bending being required. The brackets were long enough, 37 mm to take the butting up of separate lengths of the steel strip, which had been supplied in 2 m (6 feet 6 inches) lengths. Where necessary, i.e. at joints of the rail, Allen grubscrews were used to hold the two ends firmly in place and therefore avoiding the weight of the dome springingα them out when being rotated.

Now was the time for the first inspection and it was with some trepidation that I waited for the building inspector to arrive. He told me that the local authority had never had a building of this type to inspect before but after a good look round he confirmed that it met the regulations and I could now proceed with the erection of the dome.

Figure 12.5. Rail and wheel brackets.

With the success of the 24-sided walls I decided to make the dome support ring twelve sided using just over 1 m (39 inch) lengths of 5 × 10 cm (2 × 4 inch) timber with overlapping joints which were to be glued and coach-bolted together. The timber was cut to size and the joints made but before they were glued and bolted together I laid each length on the rail using steel dowels through the bolt holes to ensure that all the lengths when so fastened made an approximate circle of the same diameter as the rail. When this was done each joint was then glued and bolted in situ to keep the circular alignment. The wheels were cast in aluminium, six in all, of 70 mm (2.8 inch) diameter, and were machined with a very deep groove to avoid them lifting off the rail in the event of any unevenness, and bored 12 mm ($\frac{1}{2}$ inch) to take a steel axle. The brackets for fastening the wheels to the underside of the dome support ring were also cast and machined to take ball-races of 12 mm ($\frac{1}{2}$ inch) bore for the wheel axles. The brackets were bolted from the top of the ring to the wooden ring. Then came the test: would it rotate freely and easily? The answer was yes, so all my careful measuring and positioning had paid dividends.

Having now got a rigid and very firm base to work with I started on the struts for the dome slit which were made from 40 × 8 mm (1.6 × $\frac{3}{8}$ inch) steel flat material, the slit being 1 m (39 inches) wide. The steel strip was bent to shape against a former cut from 18-mm ($\frac{3}{4}$-inch) plywood to the correct radius for the dome and was fastened to the dome ring with bolted brackets.

The side struts are made from 3 × 25 mm ($\frac{1}{8}$ × 1 inch) steel strip bolted to the joints in the dome ring and the steel slit strip at the top. The same piece of plywood was used to make certain they took the correct shape before being bolted in situ at the top.

Having given a lot of thought to what to use to cover the dome with, I had finally settled on 1.2 mm (0.05 inch) thick galvanized steel sheet. This was on the grounds that it was rustproof, three times the weight of a similar thickness of aluminium sheet, rather cheaper than aluminium and would add a greater degree of rigidity to the hemisphere. It was only available in 2.5 × 1.5 m (8 × 5 feet) sheets, which caused some handling problems as each sheet was quite heavy and the edges were very sharp gloves became the order of the day when handling it. The procedure I adopted was to lightly pop-rivet a sheet to one strut, making certain the bottom was level and marking off the centre-point of the next strut and cutting to shape before pop-riveting it firmly in place on both struts. This proceeded all the way round. A smaller second row was necessary at the top of the dome, as the individual sheets were not large enough to cover the whole area between two struts. The inside joints were sealed with mastic and the exterior ones with a fibreglass strip for cosmetic purposes.

The slit cover, riding on rollers fastened to its underneath, was made from more of the 40×8 mm ($1.6 \times \frac{3}{8}$ inch) steel strip. This was accurately bent to shape and covered with 1.2 mm (0.05 inch) galvanized steel sheet. The cover moved freely enough but was very noisy in operation so I slipped a hard rubber sleeve on the rollers, which quietened things down substantially. I fitted an internal locking bar which meant that I could firmly lock the cover in any position from fully closed to fully open. When fully open the 37 cm telescope can view objects 100 degrees above the horizon without any vignetting.

With the dome now complete I was very pleased to find that it still rotated very freely with no tendency for it to derail at all. However, it was very noisy when being rotated and I fitted wooden blocks between the dome support ring and the outer skin which did reduce the noise considerably by reducing the area of the skin that could vibrate. One does not go looking for problems with neighbours when using the observatory at 2 am! (They are already convinced of my eccentricity.)

The decision had already been taken that the new German style mounting I intended to make would require a substantial flanged cast iron pier of about 1 m (39 inches) length by at least 25 cm (10 inches) diameter with flanges of 30 cm (1 foot) or so. Scouring of local scrap merchants produced exactly what I was looking for at the princely sum of £1.

A large block of reinforced concrete was required to be let into the ground to form a heavy and stable base for the cast iron pier so I started off by digging a 0.4 m (16 inch) square hole in the centre of the building going to 0.6 m (2 feet) square at the bottom by 0.6 m (2 feet) deep. Six lengths of 12 mm ($\frac{1}{2}$ inch) threaded steel studding were put in situ to match the holes in the flange of the pier and the cement was allowed to cure. When this had happened the pier was firmly bolted in place and levelled to a fair degree of accuracy. The final levelling of the mount would be taken care of by adjustments in the mounting itself.

Only two more stages now were left: the supply of electricity and the cementing of the floor. Electricity was laid on, taking an extension spur from my existing house ring main and running through a low current circuit breaker before powering any equipment in the observatory to assure complete safety in an outside environment.

Wooden strips were placed around the central concrete block and pier before the floor was laid to make certain that no vibration from the floor would be passed to the telescope. The floor was laid some 75 mm (3 inches) thick and levelled off. When set the wooden strips were removed thus isolating the telescope.

It was now time for the final inspection. Imagine my chagrin when the building inspector arrived with an Inland Revenue tax inspector in tow! The building passed the test with flying colours but because it was a permanent structure the Inland Revenue added £5 to the rateable value of my property. Not sufficient for me to get upset about but still a bit of a niggle.

After almost three months of work, mainly at weekends, I was there and now I could get on with the job of casting and machining the components for the new mounting which took a further three months.

With a tube length of just over 2 m for the 37 cm telescope I had deliberately designed the observatory to comfortably take a tube length of up to 3 m should

Figure 12.6. Door frame and wall construction.

aperture fever become rampant again at some time in the future.

The benefits of having a permanently housed telescope ready for virtually instantaneous use were as I had hoped for and my photography proceeded with very much greater ease.

In Retrospect

At the time of writing (December 2000) the observatory has stood the test of time well, being operational for over 33 years without much maintenance being necessary, apart from occasional painting of the wooden dome ring and the door. The odd speck of rust that has appeared on the galvanized sheet has been emeried back to bare metal and painted with an aluminium primer which matches the galvanized colour very well.

In 1995 I felt the need for a more rigid mounting and replaced the German one with a much heftier English cross-axis one supported on brick piers. I could only do this because the original design of the observatory was that much oversize for the 37 cm.

Thirty-five mm photography of the Moon and planets was superseded in 1998 by the acquisition of my first CCD camera, closely followed by my second in early 1999, so impressed was I with the huge superiority of digital imaging over 35 mm photography. Again the oversized observatory was able to accommodate a

Figure 12.7. The 37 cm telescope on the English mount built in 1995.

Figure 12.8.
Installation of the personal computer in 1998.

permanently housed personal computer to control the cameras and save the images for subsequent processing indoors without in any way restricting access to the telescope

Since 1998 the observatory has had more use than at any time during its long life and has completely vindicated my building it in the first place. The only problem that has arisen is a fir tree in my next-door neighbours' garden, which in 1966 was only 2 m (6 feet) high and did not cause any light blockage but has now sprouted to about 10 m (32 feet) and completely blocks my view from east round to south-east. Luckily I can still view objects from two hours east of the meridian though.

Were I starting again I would not change any part of the design or construction for it has truly proved its worth. I would, however, move the site from the corner of my back garden to a more central position along the back wall to avoid the fir tree blocking my view of the eastern sky.

Of one fact I am quite certain: if I had not built the observatory, I would still not be following my hobby so avidly and easily and may well have become an armchair astronomer with my increasing years!

Chapter 13

Ptolemy's Café

Bill Arnett

Background

Like so many amateur astronomers, I've always wanted to have my own permanent observatory. Of course, we would all like to have one on top of Mauna Kea but the real world intervenes and we must make compromises. So mine is in my backyard in the middle of the San Francisco Bay Area's huge light dome. Well actually, at the edge of it; I'm located in the hills between the ocean and the bay so there's no city to the west of me and little to the south. But my house and my neighbors' trees block some of my horizon. Not perfect but good enough.

Figure 13.1. Ptolemy's Café (north side).

Figure 13.2. The east side of the observatory and the deck between it and my house.

I knew from the beginning that I wanted a roll-off design, not a dome. I like the feeling of the open sky above me and the climate here is mild enough (rarely much below freezing) that a dome is not a requirement. But equally importantly, I wanted it to be aesthetically pleasing. After all, it dominates my backyard. So I decided on a "Japanese tea house" look and designed the rest of my landscaping to match.

There were two major tradeoffs I had to make. First, having the roll-off support structure on the north side as is traditional was out of the question. It would have been in the middle of my lawn. So it is on the west side where it is least visible from my house (and fortunately, not too bad for my neighbors, either).

Second, between the "tea house" design and the building codes and a pretty conservative structural engineer the structure got more massive than I would have liked. The engineer insisted on making it out of steel which greatly increased the cost (but at least I can use it as a shelter in an earthquake). It wasn't possible to have the whole south wall fold down (as is also commonly done) so I have to make do with looking through fold-down windows. But the posts and tops of the walls are pretty thick and obstruct more sky than I would have liked. I partially compensated for this by designing a ridiculously complicated pier which allows me to move the telescope up and down to get different angles through the windows and over the walls.

Building the observatory was part of a larger project to landscape the entire yard. Getting architects and plans and permits and contractors all arranged took over a year before we actually started the work.

Plans

The observatory is 14 feet (4.3 m) in the east-west direction and $11\frac{1}{2}$ feet (3.5 m) in the north-south direction (interior measures), and it is lined up with the true directions within a few minutes of arc. The original plan called for 12 × 10 feet (3.7 × 3 m) interior measure but by happy accident interior and exterior measurements got swapped a couple of times and I gained a few square feet. This gives me plenty of room to mount two scopes on piers, have some storage and working area and still move around. It is surrounded by a 4 foot (1.2 m) wide deck which is about 18 inches (46 cm) above the surrounding ground level. The walls are just under 7 feet (2.1 m) high with the roof open. The door is on the north side. There are fold-down windows on the other three sides.

There is one giant block of concrete beneath the floor with three sets of pier attachment bolts. Thus I can have one scope in the middle or one on each end. This pier is not in contact with the rest of the structure; I can jump up and down on the floor without affecting the scope at all.

Figure 13.3. The roof rolls off to the west.

I adapted the rolling roof design from one I found on the Internet. There are four 6-inch (15 cm) diameter cast iron V-groove wheels on each side.

Figure 13.4. A boy and his pier. The concrete block I'm standing on is connected to a concrete cylinder big enough around for me to fit inside and twice my height!

The wheels run on steel angles welded to the top of the main structural supports. Each wheel has a 1000 lb (450 kg) load rating. This is not as much overkill as it seems; the roof is *heavy*.

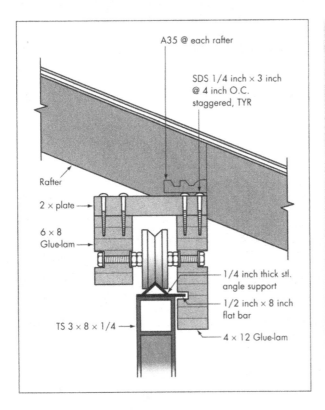

A35 @ each rafter

SDS 1/4 inch × 3 inch @ 4 inch O.C. staggered, TYR

Rafter

2 × plate

6 × 8 Glue-lam

1/4 inch thick stl. angle support

1/2 inch × 8 inch flat bar

TS 3 × 8 × 1/4

4 × 12 Glue-lam

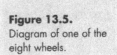

Figure 13.5. Diagram of one of the eight wheels.

I did a lot of thinking about the opening mechanism. I had always wanted a motorized one and when the roof was completed it was clear that a motor was a requirement, not an option. The roof rolls nicely and quietly but it takes a lot of effort to get it moving. I had an old garage door opener lying around so I hooked it up with a temporary kludge and much to my surprise it moved the roof easily.

I didn't want to have chain go through the wall and be exposed to the weather. So I spliced the chain to some plastic-coated steel aircraft cable. The idea is that if the distance the roof has to travel (17 feet; 5.18 m) is less than the distance from the sprocket to the point where the cable/chain must go through the wall then only cable need pass through. This meant folding the chain/cable path once but that turned out well anyway since it put the opener itself down on the floor where it's easy to work with. The chain/cable passes over the motor sprocket, up the wall and over a spare wheel to make the 90 degree turn, along the entire length of the north wall to an idler pulley attached to the far western support post where it makes the U-turn for the return path.

Attaching the cable to the roof was also a little tricky. If it was attached solidly then the roof's inertia would put a huge shock load on the motor and its gears. But if I simply put a spring in the middle of the cable then it wasn't possible to get the chain/cable tight enough to prevent it from jumping off the motor's sprocket. The answer is this: the chain/cable itself makes a complete loop with no springs so it can be tightened (via a turnbuckle at the western idler pulley) but it is attached to the roof with springs. (This is similar to the way the opener was originally intended to attach to a garage door but with much heavier springs.) It also seems to work better with compression springs rather than the more obvious extension ones; with compression springs there's a limit to the amount of stretch and so the spring doesn't get stretched beyond its elastic limit. The result works quite well. It stays tight and runs smoothly but springs three or four inches on starting and stopping with no apparent stress on the motor.

My LX200 sits on top of a telescoping steel pier which sits on top of the concrete pier beneath the floor. The idea is to be able to move the scope up and down to get the optimum angle through or over the windows for objects near the horizon. It's all steel, 10 inches (25 cm) inside diameter for the bottom section, 10 inches outside diameter for the top section. The two parts are

Figure 13.6.
Hydraulically operated
telescoping pier.

normally clamped together with a massive steel collar. The top section sits on a long-throw hydraulic jack inside the bottom section which is operated by an external hand pump mounted on the wall. I can loosen the collar, pump it up (or let it down) as much as 20 inches (50 cm), and retighten the collar in less than a minute. Were I taking long-exposure photographs I would have to redo my polar alignment but for visual work it stays aligned well enough to not matter.

Construction

Building the observatory was too big of a job for me alone; I hired a contractor. The project took more than 10 months from the first day of digging to the installation of the last cabinet. That seems like a long time but it would never have happened at all if I had tried to do it myself!

Eight 16-inch (41 cm) diameter holes and two 24-inch (61 cm) diameter holes were drilled 8 feet (2.4 m) deep and filled with reinforced concrete to support the steel frame of the building. Plus we dug an additional 24-inch (61 cm) hole 10 feet (3 m) deep for the telescope pier itself. Our "soil" is very rocky; these

Figure 13.7. Digging the foundation piers required this massive machine. It had to pass between my nerghbour's house and my gas meter with only a couple inches of clearance on each side.

11 holes took a day and half for a giant (and expensive) drill rig mounted on a huge tracked vehicle.

A steel reinforcing bar cage was built for each hole, then lowered in place and supported a few inches off the ground before the concrete was poured. The scope pier support is a roughly T-shaped block consisting of the 24-inch (61 cm) diameter 10 feet (3 m) deep base and a rectangular block on top (all connected with more rebar); it's about 4 yards (3.1 m^3) of concrete altogether. It has three sets of three 18-inch (46 cm) long, 1-inch (25 mm) diameter steel bolts for attaching the steel telescope support pier. These bolts are mostly embedded in the concrete with big washers at the bottom to help hold them in place. I held each set of bolts in place with a piece of plywood while the concrete set, and I saved the plywood to use as a template for the bottom of the steel pier which was built many months later.

The structural strength of the building is provided by eight 3-inch (76 mm) square steel tubes bolted onto the

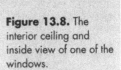

Figure 13.8. The interior ceiling and inside view of one of the windows.

concrete piers and encased and connected by reinforced concrete grade beams. Two more steel tubes hold up the roll-off support rails. All these are connected at the top by more 3-inch tubes and welded to form a complete steel box frame. But even with all that steel the structure would vibrate quite a bit when given a good kick; it got much stiffer when the wooden sheer panels were added.

A flat plate, and a 90 degree angle set on top, form the track on which the wheels run. The flat plate fits in a slot in the outside roof beam to prevent the roof from going up in a strong wind. (I'm not sure the "lift-off protection" was necessary but we do get pretty strong winds here and this way I don't have to worry.)

2 × 12 inch (5 × 30 cm) pressure treated floor joists on 16-inch (41 cm) centers sit on top of the grade beams. $\frac{3}{4}$-inch (19 mm) plywood on top of the joists forms the

floor. Several hatches were cut in the plywood to provide access to the space below the floor for fiddling with wires, etc.

The main rolling roof structure is made of six huge laminated and glued beams, two on each side to hold the wheels and one on each end to form a box. The roof itself sits on top of the box and consists of exposed redwood joists, redwood "V-Rustic" ceiling planks, plywood for structural strength and thick cedar shingles on top. The whole structure is very rigid and very heavy but it still moves smoothly and quietly.

For the walls between the steel structural members we used conventional stick framing. The inside of the wall is more redwood to match the ceiling; the outside is plain plywood for sheer strength covered with redwood lattice for appearance. No insulation! The windows are custom made individual pane panels made in the style of a Japanese shoji screen. They're sand blasted for a frosted look from the outside and hinged at the bottom to fold out and down. Each one is held closed by a simple pair of hooks and eyes. The door on the north side is a pair of sliding panels identical in design to the windows. They slide inside a round opening; in Japanese architecture this is sometimes called a "Moon door", which I thought was fitting as my main interest these days is lunar observing.

Figure 13.9. Looking south-east from underneath the open roof with all the windows open.

Figure 13.10. The north-west corner with built-in desk and bookcase. The north-east corner has another bookcase and a cabinet holding the roof-opening mechanism.

The finishing touches include carpeting, custom cabinets, desk and bookcases, phone and Ethernet wiring and a special heated case for eyepieces.

It took 11 months and 6 days from when we started digging until we were finally done; a long time but I'm very happy with the way it came out!

Lessons Learned: What I Would Do Differently Next Time

I had to make the roof roll off to the west to avoid messing up my yard too much. But this is a more serious compromise than I thought. I'm losing too much of the western sky, which is my darkest (though the seeing is usually poor that low). This problem is especially severe for the western pier location. There are a couple of

Figure 13.11. View from inside toward the south-east.

possible solutions other than the obvious rolling off to the north. First, would be to make the roof peak lower. Second, would be to make the rails longer so the roof could be pushed farther away. Third would be to have roll-off rails on the eastern side, too. This never occurred to me until long after it was too late; I had just assumed that they would be way too ugly. But not so! They actually look kind of neat (in my humble opinion).

The high walls are much more of a problem for a Dobsonian than for a Schmidt–Cassegrain or a refractor. I designed it with my LX200 in mind. But when I try to use my 10-inch, f/6 Dobsonian inside the observatory its low altitude axis means that the top of the walls is much higher in the sky. An equatorially mounted Newtonian would either have the same problem or would necessitate the use of a ladder to reach the eyepiece. A really big Dobsonian would also need more floor space. Looks like my next scope will have to be a refractor.

We get a lot of wind here, and as I was raised in San Diego, 50 degrees F and a 15 mph wind seems like an arctic blizzard to me. The walls, high though they are, don't protect from the wind as much as I would like. A dome looks a lot better when I'm shivering and going back in for my ski suit in May! A dome was out of the question this time but next time I think it might be the way to go. Or possibly a split roll-off (in which the roof rolls two ways, joining in the middle with the possibility of each half going past center) so that the opening could be adjusted to fit the weather.

My concrete pier is unnecessarily big. All that extra mass didn't seem to have any effect. Strength is not much of an issue with amateur scopes. What is important is vibration damping and concrete isn't good at that. My huge pier is not harmful but it's not helping much, either.

Even though it came out bigger than planned, the floor area could be a little bigger. There's no room to pass between the scope and the wall if the observing chair is in the way. Another 2 feet (60 cm) in each direction would help, assuming the two-scope layout; with just one scope in the middle I think it would be fine.

All that notwithstanding, I'm extremely happy with the end-result. My goals of having it both functional and aesthetically pleasing were both fully achieved! For more details and pictures see my WWW site at http://www.seds.org/billa/obs/obs.html.

Postscript: Why "Ptolemy's Café"?

Ptolemy, the famous Greek astronomer and geographer, is variously quoted as saying something like:

> I know that I am mortal and the creature of a day; but when I search out the massed wheeling circles of the stars, my feet no longer touch the earth, but, side by side with Zeus himself, I take my fill of ambrosia, the food of the gods.

As it turns out, Ptolemy's theory of the Solar System is wrong. But this quote still captures a lot of what I feel when I roll open my roof and spend a few quiet hours with the rest of the Universe.

Chapter 14

The Construction of Starbase Two

Paul Zelichowski

The Concept

Figure 14.1. Paul Zelichowski's Ski Dome in Kincardine, Ontario, Canada.

The concept that would become Starbase Two (affectionately known as the "Ski Dome") began forming in my mind soon after the purchase of a 10-inch Newtonian equatorial telescope. It quickly proved to be a tiresome cycle of tearing down and setting up equipment in my backyard. Since I wanted to delve into the wonderful world of imaging the cosmos, both 35 mm

film and electronically, as well as observing, a permanent observatory seemed to be in order. I live in Ontario, Canada, on the shores of Lake Huron in a small subdivision north of Kincardine. Because there are no street lights, the night skies are reasonably dark when it's clear. Therefore the logical choice for the observatory was in my backyard somewhere. Since my property backs on to a conservation area, I had a significant tree line to the south to contend with. This led to the idea of elevating the observatory some ten feet above grade level. I also wanted to attach it to the back of the house so as to be easily accessible from within.

I've always liked the aesthetics of the classic dome style observatory. Although not a conventional design for the average "do-it-yourselfer", a little ingenuity and imagination get the design functional. The hemispherical design also has some excellent benefits. It acts as both a windbreak on gusty nights as well as a light shield from neighbor's porch lights. The curved surface also minimizes wind shear, which is important since the area I live in is known as "The Land of Horizontal Snow"! The winter can produce some very nasty winds coming off the lake.

Equipment

A number of equipment changes have occurred since Starbase Two was built in the spring of 1998. Although the Meade Starfinder mount was adequate for 35 mm photography after some tweaking and tuning, it fell somewhat short of the standard necessary for serious CCD work. My present set-up involves a Losmandy G11 mount enhanced by the addition of an Astrometric Skywalker 2 retro-fit. This is a GOTO system that can point the telescope to any object in its internal database or, in tandem with a planetarium program, can be controlled by a click of a mouse. Along with a 10-inch, f/4.5 Meade Starfinder, I can also switch to a Celestron 6-inch, f/8 achromatic refractor or a 6-inch, f/6 Newtonian reflector. For imaging, I use a Canon AE1 35 mm camera, a Meade Pictor 208XT monochrome CCD camera and a Starlight Xpress MX5C one-shot color CCD camera.

Considerations

Attaching the dome to the house and elevating it presented some problems which needed addressing first. A tall, stable pier for the telescope and mount would be necessary as well as a supporting structure for the dome itself. I had a basic idea of what I wanted to do and drew a rough sketch of what I had in mind. Many years ago, I had built a stand-alone dome 8 feet (2.5 m) in diameter at grade level that housed a classic orange Celestron C11 Schmidt–Cassegrain telescope, so I'd had experience with the design and was quite happy with the results. Having settled on the basic design, it was time to get the project underway. I booked three weeks off from work and began my labor of love.

Construction Begins

The first arduous task was to excavate a hole for the pier. A hole approximately 4 feet (1.2 m) deep and 4 feet in diameter was cleared out. To make matters more difficult tree roots slowed progress somewhat, but after the first day the hole was complete and ready for the next stage of my plan.

The lower support structure would need to be in place so as to allow the installation and support of the pier. I decided a cast iron pipe would serve more favorably and be somewhat easier to deal with than a

Figure 14.2. The lower support structure and pier in place.

Figure 14.3. The mount mated to the pier with the floor of the dome partially complete.

concrete pier. I was able to procure a 12-foot (3.7 m) length of 12-inch (30 cm) diameter pipe after a short search of area merchants. The next step was to remove the end wall of the attached sunroom at the back of the house. The support structure would basically be a six-foot extension of the sunroom and the dome would then be built on top of it. Using 2 × 6 inch (5 × 15 cm) lumber for strength, the sunroom was thus extended. A portion of the structure was left incomplete to facilitate the installation of the pier. Since the pipe weighed about 450 lbs (200 kg), a half-dozen friends were called upon to lift the pipe straight up and deposit it in the hole. The pier now had to be positioned correctly and braced into place.

This meant raising the behemoth roughly 18 inches (45 cm) so that the top of pier would mate with the telescope at the correct height. Using a small hydraulic jack and some ingenuity (and extreme caution!), the

Figure 14.4. With the telescope in place, the dome was built around it.

pipe was raised to the correct level and supported with a few bricks. It was then leveled and braced into place. Satisfied that everything was in its place, concrete was ordered and poured into the hole. The concrete filled the hole as well as a 6-foot (1.8 m) square, one-foot (30 cm) thick slab at grade level. I over-engineered for the simple reason that if I ever wanted to upgrade to heavier equipment, I would be sure that the pier remained solid and stable.

With the pier and lower support structure in place, the floor of the dome proper was built, again using 2 × 6 inch (5 × 15 cm) lumber for strength. I also welded a piece of $\frac{1}{4}$-inch (6 mm) deck plate to the top of the pier, to which was welded a 32-inch (80 cm) length of 4-inch (10 cm) diameter pipe. A custom made adapter would then mate the telescope mount to the pier. Particleboard was then nailed to the floor and a small hatchway incorporated to facilitate access from the house.

At this point, I hadn't decided on the exact size of the dome. I installed the telescope so that I had a better idea of the exact dimensions required. Because of the custom nature of the project, I basically "flew by the seat of my pants" so to speak. I finally decided that a $6\frac{1}{2}$-foot (2 m) diameter dome would serve my purpose. The support wall of the dome would have to be low enough to allow the scope access to the lower part of the night sky. This height was set at 42 inches (1.05 m). The walls were made from 2 × 4 inch (5 × 10 cm) beams with an initial covering of $\frac{1}{4}$-inch (6 mm) Masonite. I changed to $\frac{3}{4}$-inch (18 mm) plywood for strength after some consideration.

The "Ski Dome"

The next time-consuming task was to begin cutting a series of plywood arcs that would become the skeleton of the dome. With jigsaw in hand, the next few days were spent cutting out 4-inch (10 cm) wide arcs, $6\frac{1}{2}$ feet

Figure 14.5. The wall and base ring in place. In the foreground is the ring for the dome. Note the golf-ball-bearings riding in the slot of the base ring.

Figure 14.6. The completed skeleton of the dome.

(2 m) in diameter, from $\frac{3}{4}$-inch (18 mm) plywood. I cut out various lengths, both half-circles and quarter-circles. The intent was to get as much as possible out of each sheet of plywood with minimal waste. Playing around with paper cut-outs helped to be as efficient as possible. I was able to get $1\frac{3}{4}$ circles from each sheet of plywood. Two full circles would be laminated together with screws. One such circle would serve as the top of the walls, the other the base of the dome.

The secret to making the dome rotate was by utilizing over a hundred golf balls, which rode in grooves, and acted as ball-bearings. One-inch wide plywood strips

Figure 14.7. A side view of the dome skeleton showing the side supporting arcs.

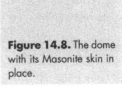

Figure 14.8. The dome with its Masonite skin in place.

were screwed to the inner and outer edges of the rings, leaving a groove in the middle, which would accommodate the golf balls. The two rings would basically be mirror images of each other. They had to be built in place, piece by piece, to ensure that the dome would rotate freely for 360 degrees. Incidentally, the golf balls I used in Starbase One were stored away (I knew they would someday be used for the same purpose!) and after some 14 years, they saw light again.

Assembly of the dome could now begin. I decided a 32-inch (80 cm) wide slot extending slightly past zenith would allow the telescope ample access to the night sky. As such, two vertical plywood arcs were cut to fit the ring 16 inches (40 cm) on either side of center. Three more support arcs on either side completed the skeleton of the dome, which was now ready for some "skin".

One-quarter inch thick Masonite was the material of choice since it's reasonably flexible and I'd had experience

Figure 14.9. In the closed position the braces tuck out of the way.

with it before in the construction of my first dome. Using cardboard templates, each pie-shaped piece of Masonite was cut out and screwed to the plywood skeleton. Once all the Masonite skin was in place, all seams received a generous amount of paintable silicone caulking. Since the Masonite needed to be sealed to protect it from the elements, I procured a couple of gallons of white rubberized paint that was used to give new life to old shingles. Several layers were then slathered over the entire dome. A fresh coat once or twice a year is required as the weather gives the dome a dingy look after a while.

The next job on the list was to make a cover for the slot opening in the dome. Initially, I fabricated a one-piece unit that slipped into the opening, but required me to go on the roof to remove. I decided it would be much easier and safer to be able to open it from inside the

Figure 14.10. The accessory tray that holds eyepieces and adapters. Note the brace holding open the lower door of the slot opening.

dome. I constructed a two-piece hinged "clam shell" type unit, which was stabilized with braces that could be clamped in any position via bolts and wing nuts.

I also incorporated a 2 × 4 foot (60 × 120 cm) "cubby hole" on the east side of the dome. It would hold a computer and provide a storage area for accessories and star charts and such. A 6-inch (15 cm) wide skirt around the base of the dome gave it that finished look.

At this point, the observatory was quite functional. Ironically, six out of seven days were clear as a bell during the construction period. I spent a few nights sitting on the roof looking at the stars, contemplating the Universe and anxiously waiting the day I could use the dome. Final touches included some interior work. I painted the inside of the dome flat black to minimize any stray or reflected light. I also finished the inside walls with drywall complete with "starry" wallpaper. Thick carpeting was also laid down, both for comfort, and, I later found, to help protect things that seemed to jump out of my hands now and then! Another necessity, of course, was electricity, so I wired in a few 120-volt receptacles. A regulated 12 V DC power supply was installed to run the telescope drive and CCD cameras.

To minimize fumbling in the dark for eyepieces and adapters, I made an accessory tray at the bottom of the slot opening. A number of $1\frac{1}{4}$-inch and 2-inch holes

Figure 14.11. The completed Ski Dome.

hold all my eyepieces, draw tubes and various adapters for imaging. Everything is within easy reach while I'm at the business end of the telescope. Another handy addition is Velcro. It's everywhere! Hand-paddles for the drive and electric focuser can be secured within easy reach when not in use. The last thing to do was to finish the entire outside of the structure with leftover siding to match the rest of the house. My labor of love was finally complete!

Conclusion

In retrospect, I should have made the dome a bit larger. While fine for one person, three people are the maximum. I did not anticipate the interest it would generate from friends, and as such have led many a group on a tour of the Ski Dome. My eventual goal is to be able to image the heavens from the comfort of a "warm room" (currently my dining room!). As such, I have a couple of computers and a laptop to run the telescope drive system, CCD cameras and planetarium software. Astronomy is probably the only science that the amateur can contribute valid scientific observations to. I hope to be able to someday contribute to the science in some small way. Perhaps a supernova search

Figure 14.12. The 6-inch refractor is ready for business.

Figure 14.13. The 10-inch reflector is ready for action.

Figure 14.14. A final shot of the closed Ski Dome.

in external galaxies would be a viable project once I become proficient in electronic imaging. As for the present, my imaging skills are slowly growing, and I'm quite content to take "pretty pictures" and observe the grandeur of the Universe and of course share it with anyone who is interested.

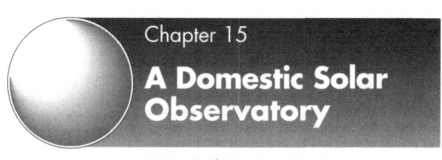

Chapter 15

A Domestic Solar Observatory

George Kolovos

Inception

Figure 15.1. George Kolovos's domestic solar observatory.

The idea of constructing a domestic solar observatory (Figure 15.1) came from a photographic exposure recorded on 9 January 1992 of the Sun's photosphere. The image revealed an unexpected phenomenon. I've been observing the Sun occasionally since 1965. My photospheric observations consisted mainly of estimations of the Wolf Number,

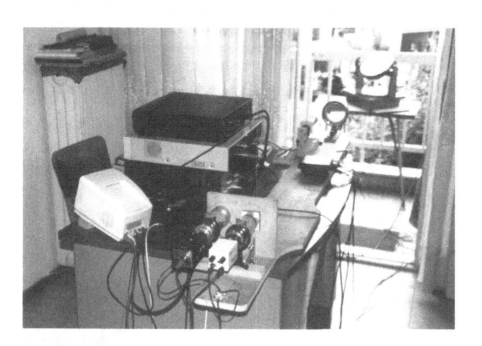

photographic exposures of the solar disk and sunspot groups.

On 9 January 1992, I took three photographs of the solar disk with a $D = 102$ mm, $F = 1000$ mm objective lens with a Barlow attached to it. In a region between two groups of sunspots located at the south-western portion of the solar disk (group nos 6993 and 6994) a filament was recorded which looked similar to those recorded through a Hα filter.

In all three photographs, the filament appeared identical, thus eliminating the possibility of a photo-graphic error. In the second exposure taken on 12h 05m 28s UT, it was found that a bright jet was visible beside the filament, starting from group no. 6996 and extending northwards (Figure 15.2). This jet was not evident in the first or last photograph taken, which was one minute before and five minutes after the second exposure, respectively. This seemed to indicate that the jet was a transient phenomenon.

I had been aware that intense flares were sometimes visible in white light. This is relatively rare but never-theless does happen. A question came to mind, could intense chromospheric filaments sometimes be visible in white light? A thorough search through the Solar Geophysical Data Prompt Reports showed that at the time when the photographs were taken there was no

Figure 15.2. The first Rapid Transient Photospheric Phenomenon (RTPP) observed on the photosphere on 9 January 1992, 12:05 UT.

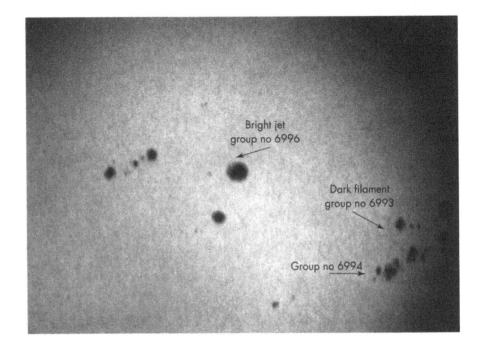

chromospheric filament between the groups of spots 6993 and 6994 nor was there any flare in the group 6996.

The events that had been photographed were clearly a photospheric phenomenon. Confident that a photographic fault was not at hand, I sent my original negatives to Mr Bruce Hardie, then director of the Solar Section of the BAA, and asked for his opinion. Mr Hardie, in a reply written on 14 April 1993, said that he himself had observed something similar on 28 July 1992 (*Solar Section Newsletter*, July 1992). With his letter he enclosed an article written by Dr T. Rackham, which was entitled *Comments on Solar Images* (Photographs 9 January 1992). Dr Rackham had examined my negatives and encouraged me to continue the observations in the same way.

This gave me the incentive to construct a permanent Solar Observatory from which I could observe the photosphere and chromosphere for several hours a day when weather conditions would allow it. Unfortunately I was not able to construct it in the countryside so I decided to build it at my home. It is located on the second floor of a four-storey building in the south-eastern suburbs of Thessaloniki, Greece. Today, I cannot state that my choice was the best one but I do feel relatively pleased by its results. Since 1994, 42% of my observations I'd classify as good, 48% as acceptable, 6% as excellent and only 4% as poor.

The room itself is oriented in such a way that on a daily basis only 4 hours of observations are permitted at one time. The instrumentation set-up is shown in Figure 15.2. They are positioned horizontally on a tabletop where they receive a (horizontal) beam of sunlight from a heliostat placed on the balcony. Below is a detail description of the instruments used to observe solar phenomena.

Heliostat

The heliostat consists of a good-quality 9-inch (22.5 cm) flat mirror. It's positioned on an alt-azimuth mount. Two motors move the mount in a horizontal and vertical direction, giving a rate of 1 turn per hour and 0.25 turn per hour, respectively (Figure 15.3). The motors are controlled automatically by the guidescope shown in Figure 15.4.

A 25 mm (1-inch) solar image is projected on to a surface where three phototransistors are positioned. A

Figure 15.3. The alt-azimuth mounted heliostat.

triple DC amplifier (Figure 15.5) uses the current from the phototransistors to give commands to the heliostat.

Cameras

The cameras are commercial black-and-white CCD video-cameras, with a chip $\frac{1}{30}$ inch wide and 0.1 lux sensitivity. The gamma factor is 0.45 or 1 which can be changed with a switch placed on the CCD (Figure 15.6).

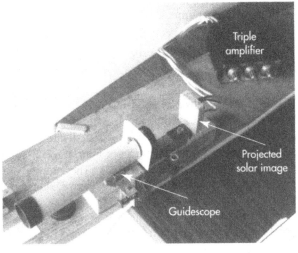

Triple amplifier

Projected solar image

Guidescope

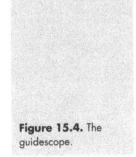

Figure 15.4. The guidescope.

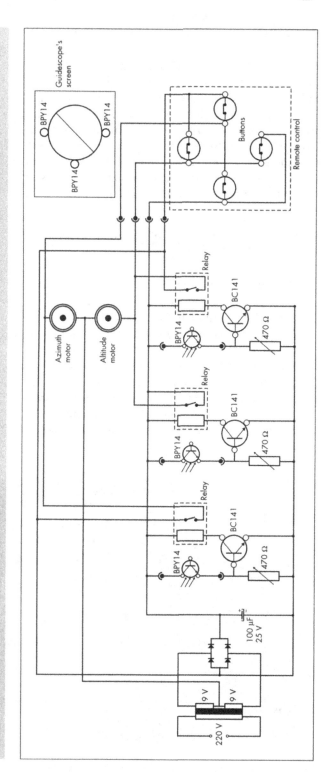

Figure 15.5. The triple DC amplifier.

Monitor

A commercial 5-inch (125 mm) monitor with high contrast allows me to observe part of the solar surface with a scale of 36250 km/cm or 5 arcsec/mm. Of course I can scan the whole disk and focus only on interesting active regions. All observed phenomena can be recorded on a video-cassette recorder, or printed with a video-printer. Optionally, I can connect the monitor to a PC equipped with a TV card, something I tested once but have not had the opportunity to use.

Filters

(a) For photospheric observations, I use a violet inteferometric filter with a peak at $\lambda = 405$ nm and a bandwidth of 16 nm. I came to this decision after testing various types of filters. The aim was to obtain a high-contrast solar image. The violet filter was finally the one that gave me the best results.

(b) For chromospheric observations the use of a Hα filter is necessary. For me it was important to see if the photospheric phenomena observed since 1992 is correlated with any other phenomena observed on the chromosphere.

Figure 15.6. The devices.

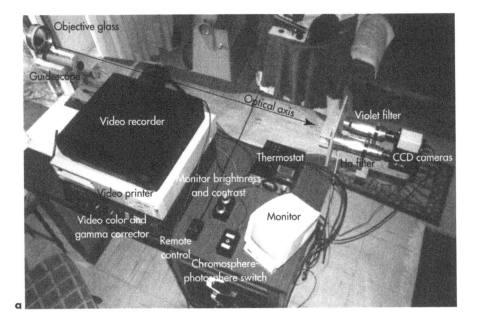

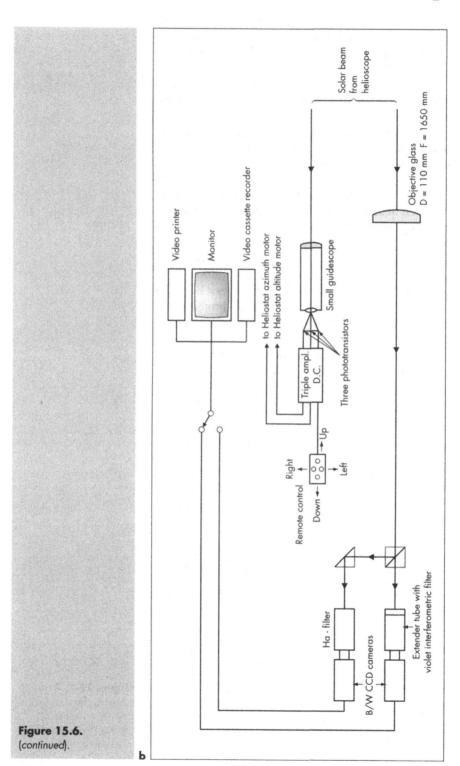

Figure 15.6.
(continued).

b

Unfortunately, a good-quality Hα filter is relatively expensive. But after having read D. Menzel's book *Our Sun*, page 151, where he quotes "The manufacture of such filters does not lie beyond the skill of an experienced amateur telescope maker", I thought it would be a good idea to try and construct one of my own.

Without knowing the outcome, I began optimistically to construct my Hα filter. I found that trying to construct the filter was very difficult. At this point I had progressed too far into the project to stop. All the necessary equipment, such as polarizers, quartz and calcite crystals, had been obtained. Also during the process, cutting and polishing machines including a spectroscope with a resolution of 1 Å/mm, were constructed.

After more than a year of hard work I succeeded in creating a Lyot-type Hα filter with a bandwidth of 1 Å. I admit that this filter is not the best, but at least it ensures that through this I can easily observe all chromospheric phenomena such as granules, filaments, flares and of course the prominences (Figure 15.7).

The structure of the filter is shown on Figure 15.8. The details of constructing such a filter is enough to fill an entire book. Below, I give the basic idea of making a home-built Hα filter, for those amateurs who enjoy constructing.

1. The main surface of the crystal plates must be absolutely parallel and contain the optical axis of every crystal.
2. All axes of the crystals must be parallel and in the same direction. The same rule must be kept for the

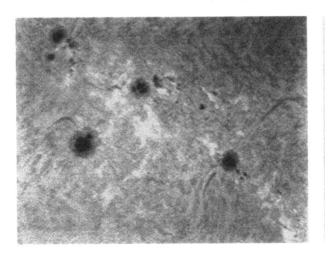

Figure 15.7. A typical chromospheric active region, observed through the home-made Hα filter taken on 19 June 2001, 12:34 UT.

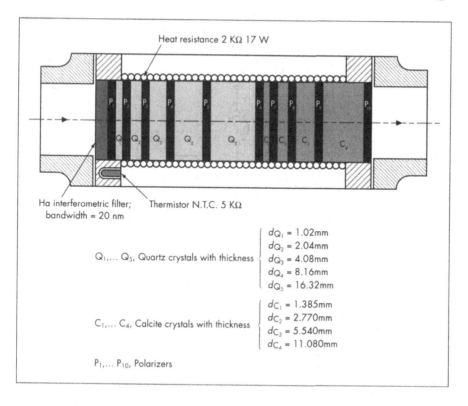

Heat resistance 2 KΩ 17 W

Ha interferometric filter; bandwidth = 20 nm

Thermistor N.T.C. 5 KΩ

$Q_1,... Q_5$, Quartz crystals with thickness
$$\begin{cases} d_{Q_1} = 1.02mm \\ d_{Q_2} = 2.04mm \\ d_{Q_3} = 4.08mm \\ d_{Q_4} = 8.16mm \\ d_{Q_5} = 16.32mm \end{cases}$$

$C_1,... C_4$, Calcite crystals with thickness
$$\begin{cases} d_{C_1} = 1.385mm \\ d_{C_2} = 2.770mm \\ d_{C_3} = 5.540mm \\ d_{C_4} = 11.080mm \end{cases}$$

$P_1,... P_{10}$, Polarizers

Figure 15.8. Detail of the Hα filter device.

axes of the polarizers. The angle between these two directions must be 45 degrees.

3. All crystals and polarizers must be stuck together with Balsam of Canada. It needs to become one solid piece and must be free of pressure. The calibration of the Ha line can be done by heating the filter at 38.6 °C. The temperature must be kept steady. This can be done with the help of a thermostat (Figure 15.6). The thermostat circuit is shown in Figure 15.9. It is a patent of my own and has been functioning for several years without any problems.

The Observations

The phenomena that have been observed since 1992 are sometimes rare. Although the method used is simple, strange phenomena are visible in both the chromosphere and photosphere and these are listed below:

1. On the solar photosphere, in addition to the already-known phenomena (spots, granules and faculae),

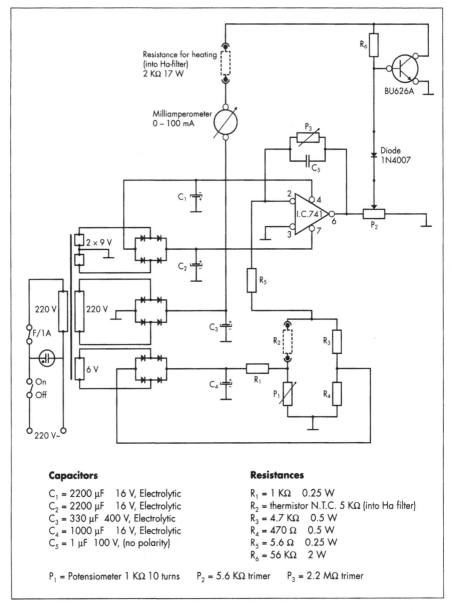

Capacitors

C_1 = 2200 μF   16 V, Electrolytic
C_2 = 2200 μF   16 V, Electrolytic
C_3 = 330 μF   400 V, Electrolytic
C_4 = 1000 μF   16 V, Electrolytic
C_5 = 1 μF   100 V, (no polarity)

Resistances

R_1 = 1 KΩ   0.25 W
R_2 = thermistor N.T.C. 5 KΩ (into Ha filter)
R_3 = 4.7 KΩ   0.5 W
R_4 = 470 Ω   0.5 W
R_5 = 5.6 Ω   0.25 W
R_6 = 56 KΩ   2 W

P_1 = Potensiometer 1 KΩ 10 turns P_2 = 5.6 KΩ trimer P_3 = 2.2 MΩ trimer

there are certain other rapid transients. The shape of the Rapid Transient Photospheric Phenomena (RTPP) – as I named them – usually look like filaments, but they can also appear as dark or bright haze spots, widened rings, loops at the edges of large H-type spots and jets emerging from the umbrae of the spots (Figure 15.10). It is strong proof that the photosphere is much more active than previously believed.

Figure 15.9. The thermostat circuit.

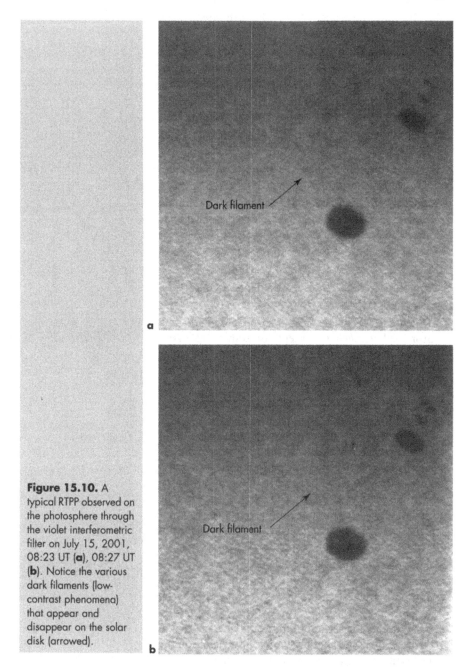

Figure 15.10. A typical RTPP observed on the photosphere through the violet interferometric filter on July 15, 2001, 08:23 UT (**a**), 08:27 UT (**b**). Notice the various dark filaments (low-contrast phenomena) that appear and disappear on the solar disk (arrowed).

All of them are low-contrast phenomena and can be observed everywhere on the solar surface, but they appear 2.5–3 times more near active regions. RTPP are absent from already-evolved active regions, where flares are the usual behaviour observed in the Hα line. References for such phenomena were found and are listed below:

Barlett, James C., Jr: 1947, A new aspect of the Sun as a variable star, pp. 367–375, Popular Astronomy, August 1947, Vol IV, No 1, Edited by Curvin H. Gingpich, USA.

Bray, R.J. and Loughhead, R.E.: 1964, Sunspots, pp. 64–69, Vol 7, The International Astrophysics Series, Chapman and Hall Ltd, 11 New Fetter Lane, London EC4.

Hardie, Bruce: The British Astronomical Association, Solar Section Newsletter, July 1992.

Flammarion, Camille: 1884, Revue d'Astronomie Populaire, p. 381–384, L'Astronomie, Publiée par Camille Flammarion, Troisième année, Paris.

Menzel, Donald H.: 1959, Our Sun, p. 122–152, Harvard University Press, Cambridge, Massachusetts.

Trombino, Donald F.: Nov. 1995, Our Discovery of a Solar Mystery, Sky & Telescope, pp. 98–100.

In general, RTPP are not correlated with any chromospheric events, except for some rare cases. Maybe these phenomena are proof that the solar photosphere is a huge magnetic carpet, something recently investigated from SOHO.

2. Oscillating dark filaments are usually observed on the solar chromosphere, mainly near active regions, which precede or follow a flare. In some rare cases the oscillating filaments can appear at regions far away from an active one. The period of the oscillation can vary from 0.1 to 1 sec and its full amplitude can reach the value of 15 arcsec! Of course, this cannot be oscillation of matter but of energy, so that I suggest it might be oscillating magnetic fields. So far, I have not been able to find any reference to this phenomenon.

3. Sometimes the spots can appear with two umbrae, possibly of opposite polarization, inside a single penumbra. In these cases, flashes can appear between the two umbrae, lasting for 0.05–0.1 sec. The flashes can be visible for several minutes in the same region (Figure 15.11). The phenomenon is also visible on the photosphere, but it is much more fascinating on the chromosphere. I found one reference to a similar phenomenon in an old book of C.A. Young, *The Sun*, p. 124, New York, 1898, P. Appleton and Company.

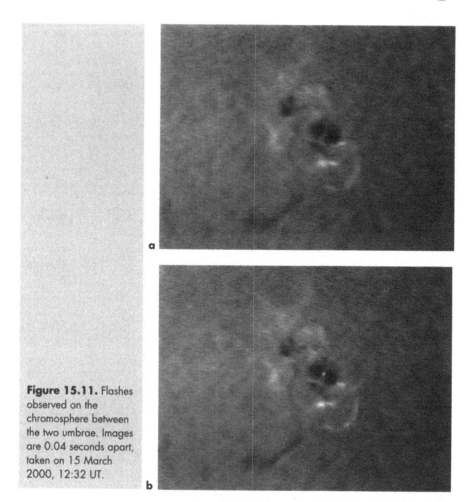

Figure 15.11. Flashes observed on the chromosphere between the two umbrae. Images are 0.04 seconds apart, taken on 15 March 2000, 12:32 UT.

Conclusions

The basic conclusion I drew, after almost 10 years of continuous observations, is that sometimes unknown or rare phenomena can be observed on the solar disk. I observe them almost every day, with simple methods and easy-to-build equipment. Observations with a CCD video-camera can derive important information about rapid transient phenomena that can be recorded in real time, something not widely known so far. Solar observation is something that everybody can do. All we need is patience and courage to keep observing as much as we can of our Sun. To date I have recorded many hours of strange and rare phenomena, most of them really interesting even for a serious professional.

Observations continue.

Chapter 16
Coddenham Observatory in Suffolk, England

Tom Boles

I had originally set my observatory up in Northamptonshire in the midlands of England. The logic of doing this was hopefully to get more cloudless nights for observing; as anyone who has lived in the United Kingdom knows, its unpredictable weather isn't one of its main tourist attractions. Soon after moving, development of commercial sites nearby generated enough light pollution to make the site unusable.

I found a reasonably dark site in Coddenham, Suffolk, which seemed near-perfect but had one main drawback. The house that I had found, which was well-situated on top of arguably the highest point in Suffolk with a clear horizon to the east and south, only had a very small plot of land attached. Inquiries at the local hostelry revealed that there was a friendly local farmer who might rent me some land and also the possibility that whoever moved into the ten-acre farmhouse on the other side might sell a small plot of land. I took the risk and moved in. This is not something that I would recommend to anyone else. Although everything is settled – I have rented a bit of a field from the farmer and later, after building my observatory, I bought an extension for my garden from my new neighbours – it was a very stressful time. On reflection it was also a very high risk, but things have worked out exactly to plan.

The Coddenham Observatory is perhaps unique in that it was designed for a specific purpose, that of discovering supernovae. To meet this requirement it had to be able to be automated so that the telescope control and image comparisons (the new patrol image against a previously stored master image) could be done indoors. After passing my fiftieth birthday I found it more and more difficult, if not dangerous, to stay out all night in January and February for fourteen hours at a time at subzero temperatures. In Northamptonshire, I had a small astronomical dome, some 10 feet (3 m) in diameter. It was ideal as the region could on occasion be very windy. The dome offered maximum protection, but it was difficult to modify for automatic patrolling. I did fit a drive motor to turn the dome so that its opening was always in line with the telescope – no mean feat after I had upgraded to a German equatorial mount. These mounts, of course, must normalize, that is, they flip over when they pass the meridian to ensure the counter-weights stay lower than the optical tube. This prevents the telescope driving itself into the supporting pillar and making one astronomer very unhappy. This put heavy demands on the control software. It did work, albeit slowly, but it was also noisy and, with neighbours close by, it was not an ideal set-up.

A roll-off roof was the answer. It would allow full access to the sky without having to resort to additional mechanisms to move it, mechanisms that could go wrong. It would of course allow me to add a second telescope. The plan was to have two automatic telescopes in the roll-off observatory and to keep the dome, but put a telescope in there with an eyepiece permanently fitted for some visual observing.

Figure 16.1.
Observatory with roof and south and east walls opened, showing the two local PCs for telescope control.

I was very lucky in having a very supportive parish council. These people play a major role in approving planning permissions. Coddenham is a protected village, in a conservation area. The full significance of this was not apparent when I moved there. Fortunately, the council decided that having a (famous?) observatory in the parish would add value to the village rather than detract from it. Planning permission finally came through, with a few caveats regarding landscaping, such as surrounding it with trees so that it could not be seen from a hill two miles away. It was difficult to convince the county council, who insisted on the landscaping, that an observatory must also be able to see out to be effective. While all this was going on, I started the construction. It was December. The first wood for the construction arrived on 23 December and it was to be one of the wettest and coldest winters in Suffolk for some time.

The Construction

Suffolk has many advantages. It also has a few disadvantages. One of these is that tradesmen seem to be trained never to say no. They promise the world and then promptly forget they ever knew you. I finally found one who delivered what he said he would. This was the task of delivering and levelling ready-mixed concrete in sufficient quantities to cover the 6×3 m (20×10 foot) base that the observatory would stand on. Not only did he agree to build the base but he also dug out and lined the two rectangular holes where the brick pillars for the mounts would later be constructed. As astronomers always, from preference, put their observatories well away from roads, access to the site was not easy.

The plan was to start building the panels for the walls while the concrete base was laid and given time to cure. This worked quite well in practice. The panels could be built in the garage. The only concern was, would the base and panels fit once finished? This concern was groundless. When I measured the finished base it was accurate to one-half of a centimetre (0.2 inches) along its 6 m (20 foot) length. The basic plan was to have only one fixed wall – the north wall, which would include the door. The other three walls would be in two parts. The top half-metre would hinge out and down to rest on

wooden brackets secured to the outside walls. This would give maximum wind protection (for a roll-off construction) yet allow the horizon to be seen on all three sides when conditions permitted.

In the UK at the time of construction, the timber industry seemed to have gone only half metric. Wood was bought in metric sizes, which approximate the old imperial (inch) sizes, but not quite. Panels were also bought in metric sizes but these seemed to be exactly the old imperial sixes (8 × 4 ft) but simply converted to imperial. This made planning the dimensions, to minimize wastage, a little tricky. Being old-fashioned I kept to the imperial sizes and only converted to metric now and then to check for possible errors.

I opted for building the panels with 4 × 2 inch (10 × 5 cm) untreated timber and put the strapping in with the traditional 16 inch spacing. Using this I could use $2\frac{1}{2}$ 8 × 4 ft (2.4 × 1.2 m) panels lengthwise on the inside of the walls. This and the wall thickness would give sufficient length to overhang the base and ensure rainwater would drain away and not collect, and so rot the wall timbers. A similar approach was taken for the small walls. Ship-slap timber strips were used on the outside. The frames were constructed first and a large 2 m (6 foot 6 inches) builder's set square used to ensure that they were square. The inside panels were added first to give rigidity and hold the frames square before the ship-slap was nailed in place. This was repeated for all eleven panels, one for the north wall, two for the south (to allow the top to hinge down) and four for each of the east and west walls.

Figure 16.2. View of the observatory looking east, with the dome of the public observatory which houses the 10-inch Schmidt–Cassegrain telescope.

The eaves for the roof were also constructed from 4 × 2 inch (10 × 5 cm) timber, the individual struts being attached to each other with nail plates. The trolley on which the roof would sit was also made indoors. This was made from 4 × 4 inch (10 × 10 cm) timber. The joints were cut but not assembled and the 5-inch (125 mm) nylon wheels, which I had managed to find and purchase over the Internet, were also fitted. The most challenging part of the construction was assembling the roof. This was done in situ. Its final weight turned out to be $1\frac{1}{2}$ tons so the option of assembling and lifting it into place was quickly dismissed.

The longer trolley sides with the wheels were positioned on top of the sidewalls on top of blocks of wood. The wooden blocks were to allow them to stand unsupported with the wheels off the walls while the shorter sides were screwed on. The next task was screwing the eaves into place and adding support struts between them at the apex to increase their rigidity. At this point the wooden blocks were removed, allowing the wheels to rest on the

Figure 16.3. The marine winch used to open the $1\frac{1}{2}$ - ton roof.

sidewalls. A length of 1-inch (25 mm) square timber the length of the long walls was put on each side of the wheels to act as guides to keeps the wheels central. (This was later to be modified.) The 4 × 4 inch (10 × 10 cm) stilts to hold the rails that the roof would eventually roll on to were hammered into Met posts that had previously been set in two concrete channels built at the same time as the base. The cross-supports and rails were added and the 1 × 1 inch (25 × 25 mm) guide channel extended out over the runners. At this stage the roof's ability to roll was tested. It worked, but it was very heavy. Wood panels were then added to the eaves and the roof was finally covered with roofing felt.

The roof was now impossible for one person to move. It was at this stage that a marine winch was added and a block-and-tackle assembly constructed from pre-stretched synthetic rope to enable the roof to be manhandled.

The Telescopes

The telescopes are two C14 optical tube assemblies from Celestron. They are mounted on two Paramount German equatorial mounts from Software Bisque in the US. These mounts were chosen because they had features specifically relevant for automatic patrolling. The Paramount is one of the most accurate mounts available and is reasonably affordable by amateurs. Two Apogee AP7 cameras are attached to the Cassegrain focus of each telescope. The reason for selecting the AP7 is sensitivity. It has 24-micron pixels (the light-sensitive cells that record the image). This is a relatively large size for pixels, but it is their size that is one of the things that gives them their increased sensitivity. The second thing is that the electronic chips holding the pixels have been thinned by removing part of their surfaces to allow them to be reversed, so allowing the light to fall on their rear surfaces, thus increasing sensitivity and spectral response (back illuminated chips). This increased sensitivity makes them highly suitable for searching for faint supernovae. When conditions permit, magnitude 19 stars can be recorded in only 60 seconds. To complete the ability to fully control the telescopes remotely, each is fitted with an electric focuser that can be driven by software via power from the mount. Both the cameras and the mounts require control software to allow their remote and automatic operation.

Figure 16.4. The two C14 Schmidt–Cassegrain telescopes on robotic Paramount German equatorial mounts.

The Control Software

Four pieces of software work together and achieve the required results. A scripting program, into which the user writes instructions, controls the other elements. This will typically contain the astronomical target (galaxy), the integration (exposure) time for the camera and any delay that is necessary in order to allow the telescope to settle after each automatic slew. The scripting module controls the camera control program, the module containing the galaxy positions and telescope control algorithm. The suite of programs has a fourth module, which sits between the scripting program and the telescope control. This modifies the target sky position for each galaxy, depending where in the sky the telescope will be pointing. This is necessary because of errors in pointing that can occur because of flexure in the mount itself, poor polar alignment and even atmospheric refraction. For this to work effectively, a few hours must be spent to train the telescope and mount assembly. The software remembers where corrections are manually made during the training run and makes the same corrections automatically from then on. All the software mentioned above resides on two PCs, which sit inside the observatory building and are permanently connected to the mount and camera.

These two PCs provide the automatic control of the telescopes.

Remotely Controlling the Telescopes

Once scripts are loaded, the telescopes will continue to slew, image and save on disk images of the selected galaxies until the script is completed or until the operator intervenes, probably to re-focus or more often due to weather conditions. Access to the PCs' keyboards is necessary to make these changes. In order to accomplish this without being in the observatory a Local Area Network (LAN) has been installed connecting the two PCs in the observatory with control PCs in the warmth of the house. One room is put aside as a "warm room" for accessing the equipment from indoors. The LAN basically connects the two computers in the observatory together and allows them to communicate with the computers indoors over a single cable. The software drivers to do this are contained within the Windows operating system. To take control of the observatory computers from the house requires an additional program. The one I use is called pcAnywhere and is commercially available from Symantec. This program allows the keyboard and mouse of a second, guest computer (indoors) to take

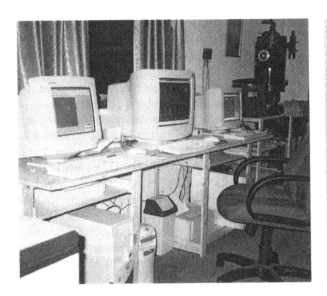

Figure 16.5. The warm room – the luxury of remote control: two PCs for telescope control; one PC for image comparison.

control of the keyboard and mouse of a host computer (in the observatory). Everything that appears on the host monitor is seen on the monitor indoors. It effectively moves the observatory PCs indoors and permits the operator to work in comfort.

After the telescopes are brought on line, which must be done from the observatory, scripts can be loaded, changed, halted and the telescopes can be focused from indoors. New directories can be created, the operating temperature of the cameras altered, the quality of the images monitored (for changes in the weather, or focusing) and images sent from the disks on the observatory computers indoors. This allows searching for supernovae to start while the telescopes are still patrolling. At the end of the night's observing, the camera coolers can be switched off and the mounts parked in their home position so that the roof can be rolled back to its home position.

The Second Observatory – The Dome

There is little I can claim to have contributed to the construction of the second observatory. The dome is a commercial product. My input has been limited to disassembly and reassembly when I moved from Northamptonshire to Suffolk. My next project is to refit the drive system to rotate the dome. I will however need to improve on the external drive housing to improve its susceptibility to rain that was experienced when it was last fitted. I have fitted a radio security system to it that works very well. This was purchased from the local DIY superstore. I have extended this to protect the main building as well.

What Would I Do Differently?

The wooden guides I have already mentioned. The guides worked fine but I had little confidence in them given the final weight of the roof. I changed them for an aluminium track, which was much stronger and protected the wooden surface of the walls from

Figure 16.6. Sir Patrick Moore and the author at the official opening ceremony on 19 August 2001. Photo by kind permission of Martin Mobberley.

developing ruts under the wheels. I made three walls fold down, the east, south and west. With experience I wouldn't have designed the west wall to open, given the access to the horizon on the west side being slightly blocked by the top of my house roof. When I fitted the aluminium runner on the west side, I did it in one complete track. This gives additional strength to that wall and makes the roof easier to open. It, of course, prevents the wall from opening. If I ever change my mind, all I need do is to cut through the aluminium to have an opening wall once more. All in all, the observatory is working fine and it continues to fulfil its purpose, that of regularly discovering supernovae.

Chapter 17
Building The Crendon Observatory

Gordon Rogers

History

Figure 17.1. The Crendon Observatory.

Twenty years ago I bought a 3-inch Zeiss refractor in a photo-shop sale. It stood in the conservatory for years: on warm nights it would occasionally get an airing when the Moon or planets were well placed. On 13 June 1994 I was observing the new moon with 30 degree terminator but

the optics suffered from three smudges of "condensation". In ten years I had never had a problem with condensation and it was a dry night after a hot day. Nevertheless, wiping the exposed optics, the "problem" persisted so it had to be internal. Being an undriven telescope the Moon drifted across the eyepiece and the "blobs" moved in synchronization with the Moon: this had to be cloud but there was not a cloud in the sky. As the sky darkened I could see more. One of the blobs had a stem which issued from a small crater (perhaps Santbech). The stem was milk chocolate in color, lightening as it rose to a white cloud. This was so dramatic that I got my wife and three neighbors to look and drew what I could see. In my determination to speak with an astronomer there and then I made fourteen telephone calls: unable to get through to an astronomer in this country I phoned NASA and got the answerphone: the last four calls were to Hawaii where I finally made contact with a lunar astronomer who would look at the Moon that night (clouded out). Subsequently various experts have concluded that I was seeing things. The man running the Clementine mission at the Moon at that time thought I may have seen a spacecraft venting gas!

Since I had no means to photograph what we saw I determined to update to something that would permit photography. This was an 8-inch Celestron Schmidt–Cassegrain with which I managed some decent pictures of comet Hyakutake but had to set the telescope up and polar align it each evening on the terrace. Why not go mad and buy a 16-inch alt-azimuth Meade LX200 and put it in a garden shed with a roll-off roof? This was a tremendous step forward but if the night was not set 100% fair there was the problem of closing the roof and re-opening. Each time this was done I certainly lost focus and often lost synchronization. Why not treat yourself to a proper dome? The search was on to see if any location could sustain an observatory.

Whereabouts

I live in a house on a slope at the edge of the village. The original building is thatched and there have been several extensions. There are mature trees on three sides so that an observatory would have to be elevated:

I also saw it as fundamental that the observatory should be accessible directly from the house. There was only one location where the structure could go and this would entail: (1) building on what I knew from previous experience was bad ground, and (2) buying a narrow strip of land from a neighbor.

I knew that technically it should be possible to overcome the ground conditions but could I buy the strip of land? It was owned by a close friend and neighbor and the land was totally useless to him – sounds easy. Hang on – Richard is as generous a "mine host" as you will ever find when he is at home. In business he is a tough dealer (he is also a neighbor who shared my sighting of the event at the Moon). After months of negotiations I owned the land and Richard was building a new wing on his villa in Spain! Having got this far what sort of dome from the great variety that are out there?

Dome

After much reading of books I concluded that the overriding requirement was that, being stuck high up in the air on a very exposed slope, it must be robust. I looked at some fibreglass domes in America but eventually decided on a steel Ash Dome. Now I have to get permission to build it.

Planning

Not easy! Being hard on the edge of the village the planning officer did not like a three-storey structure culminating in a shiny dome adjoining a cottage with a thatched roof. At the time I wanted to submit the planning application there was a big fuss about a mobile phone mast that was to be erected not too far from the village school. The local MP was involved and clearly anything "high tech" was going to have a rough ride in the village. However, thanks to very robust support from near neighbors, permission was eventually obtained after some months of negotiation. Armed with the magic piece of paper how do you get it built?

Construction

What Builder?

If I had gone to Yellow Pages and asked ten builders to quickly erect a three storey observatory that was deliberately out of square on bad ground with a circular upstand to be let into a conventional tiled roof I think I know the ten answers I would have got. Fortunately I have a builder son, Chris. Call it nepotism but I think he is good. I put him on the spot with the blood angle and it was my good fortune that he was just finishing one job and waiting for consent on the next.

Ground Conditions

You cannot fool around with telescope pier foundations. Step one, get expert consultants to thrust bore to eleven metres, as shown in Figure 17.2, taking regular samples for laboratory analysis. Step two, arrange for expert contractors to design and construct an appropriate footing. This turned out to be three 8-metre (26 feet) deep 9-inch (22.5 cm) reinforced piles topped with a reinforced slab as illustrated in Figure 17.3. Needless to say, the piling operation punctured the drains and Figure 17.4 shows the repair job.

Figure 17.2. The rust sampling.

Figure 17.3. Piling.

Figure 17.4. Drain repair.

The Pier

How do you construct a $2\frac{1}{2}$-foot (90 cm) concrete pier 18 feet (5.5 m) vertically? We discussed many alternatives but eventually settled on a one-piece sewer pipe made of plastic but with heavy ribs to give strength. Figure 17.5 shows it arriving. The only way to position this weighty object over the steel reinforcing was by crane. We very much wanted to cast the pier in one piece but were concerned about containing the concrete at the foot of the pipe – eight tons of "going

Figure 17.5. The pier tube.

off' concrete outside the kitchen window would not endear me further to my wife Margaret who is not entirely enamoured of the whole enterprise! Firstly the base of the pier was built round and concreted externally and left to cure for several days. Then the big pour was started to about one-quarter height. A sample of the concrete was held back and when this started to stiffen stage two was executed, and so on. The pier stood – hooray!

General Building

I doubt that Chris would call any part of the work general building since the structure is not rectangular:

Figure 17.6. Steel joists.

Figure 17.7. The ring beam.

Figure 17.8. The carcass.

virtually every piece of carcass timber had to be individually measured, there were all sorts of irregular abutments, a steel dome ring to design and support, much lead-work where, again, each piece had to be tailor made and a pitch finish flat roof that somehow had to be made weather-tight. The building was timber-framed for least heat retention with a steel frame for rigidity: Figure 17.6 shows the joists arriving whilst Figure 17.7 shows the base ring beam and Figure 17.8 the general carcass. There is also the building inspector.

He scratches his head a bit but generally approves of what is going on. An astronomical dome is a very specialist construction and in every amateur observatory that I have entered you have had to duck. With great ingenuity Chris came up with a scheme for a staircase that would give easy access. I am 6 foot and would have had clear headroom but two metres (6 foot 6 inches) was required and there would be no let-up on this. Result: access is by ladder – well done.

The Dome

It took two months to carry out the groundwork and three months to get the building ready to receive the dome. Amazingly the container from Illinois arrived on time, five months after being ordered, just as it was needed even though it had been held up by a hurricane and Labor Day. Chris has a number-one man, Terry. We opened up the container and I shall never forget the look on Terry's face (Figure 17.9) when told he had to put it together. It looked so daunting with stacks of bars, sheets, brackets, bolts, unctions and boxes, together with a construction manual running to 76 pages. Figure 17.10 shows the shutter in transit. Building this would be bad enough at ground level let alone 16 feet (5 m) high. The makers had gone to great pains to ease the construction process. The dome had been assembled at the works prior to dispatch so that

Figure 17.9. The container arrives.

Figure 17.10. Shutter to site.

you knew that holes must align. Parts were very carefully labelled and the instructions thorough. The second most difficult area was getting the baseplate precisely level: with this done, good speed was then made. The most satisfying part was inserting the dome panels. This was done on a warm and sunny day. The instructions were to slide them together with the help of washing-up liquid – it worked and we had a dome (Figure 17.11). The worst part of the job was installing the shutter. This was on a dark and showery day and a crane was needed. Chris and Terry were inside to instal

Figure 17.11. Chris at work.

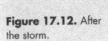

Figure 17.12. After the storm.

the rolling gear and I was outside. Just at the wrong moment a squall came through and the force of the wind snatched the shutter and spun it like a top. (Figure 17.12 gives a feel for the occasion.) The crane driver had good reactions and quickly lifted it away, returning the piece when it had calmed down. I was greatly relieved on hearing that the shutter was secured.

Figure 17.13 is of Simon turning his hand to tiling: in one day he mixed eight tons of concrete for the pier – not a man to take on in a brawl!

With so many abutments, a circular upstand through a tiled roof and an extensive flat roof, so carefully built by Terry (Figure 17.14), I braced myself for water entry on the first storm. A few drops did come in but only from the shutter – a little extra mastic soon made the whole structure watertight.

Figure 17.13. Simon tiles.

Figure 17.14. Terry roofs.

Conclusions

It was worth all the effort. In selecting the shutter assembly I am pleased with the choice I made. I opted for the wider shutter with a lanphier window which can be attached to a metallic screen so that weatherproof observing is possible. This is good for showing friends planets on a cold and windy night but I do not propose a layer of glass for imaging. The lanphier window can be parked or hauled up with the main shutter. This can slow observing down for a few minutes when needing to close or open the main shutter in order to pick up, or park, the window. Having the computer room immediately below the telescope is a boon and robotic focusing has taken away a big problem area.

One unexpected reaction is from the general public. Very much interest is expressed and many want to have a look. The farrier putting up a grating just had to see how it worked, as did the Rastafarian Omega driver, the Dutch actress, the chimney sweep and especially young children.

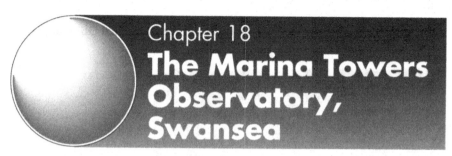

Chapter 18
The Marina Towers Observatory, Swansea

S.J. Wainwright

Preamble

This chapter is not about how to do it, but is rather about how it was done. It is the story of a small public observatory located in Swansea, South Wales, in the United Kingdom. It tells how a local astronomical society and local government were able to get together to produce an astronomical facility where there could have been a folly.

The Marina Towers Observatory stands at the end of the promenade in the Swansea Maritime Quarter, one of the most prestigious developments in South Wales. It is a magnificent piece of architecture standing as a pair of towers, which from a distance resembles a spaceship poised for launch (Figure 18.1). At the edge of the sea, the Marina Towers' unique 0.5 m Shafer–Maksutov telescope (Figure 18.2) cuts through the City's light pollution effortlessly, to reveal exciting views of the Moon and planets, delighting and edifying astronomers and members of the public alike.

The story of this small public observatory can be a source of encouragement and inspiration for any astronomical society contemplating the formation of a working relationship with local government with the aim of building and running a small astronomical observatory for the benefit of all.

Figure 18.1. The Tower of the Ecliptic, The Marina Towers Observatory.

How it all Started

The story begins in 1984, the year that Big Brother proposed the building of an observation tower on the sea front in Swansea's Maritime Quarter.

The Swansea Astronomical Society already had an astronomical observatory situated on the university playing fields at Fairwood Common on the Gower peninsula. At that time the Fairwood Observatory housed an ageing 9-inch Newtonian telescope dating from 1949. It was time for change should the opportunity arise.

The decision of the Local Authority to build an observation tower on the beach was noticed by the secretary of the Swansea Astronomical Society, the Late Gerry Lacey. The proposal for an observation tower was a concept that involved the possibility of placing a "penny in the slot" type telescope at the top of the tower that would allow members of the public who

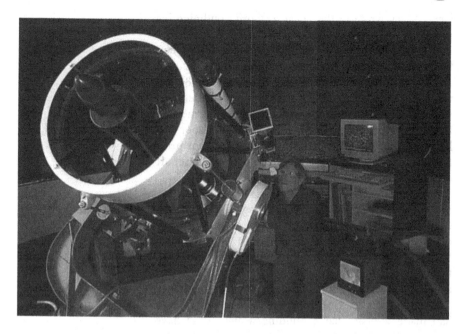

Figure 18.2. The 0.5 m Shafer–Maksutov being operated by Senior Observer Fred Whittle.

climbed its staircase to enjoy views over Swansea Bay towards Mumbles Lighthouse in the distance.

Gerry Lacey saw in this proposal an opportunity for the Swansea Astronomical Society to work with the Local Authority to develop the Observation Tower as an Astronomical Observatory, which would confer considerable added value for the functioning of the building in the context of the local public and of tourist visitors to the city.

Gerry's imagination and vision were eventually to fire the enthusiasm of the Local Authority. He arranged a meeting with the City Senior Architect, and the Tourism Development Officer. Gerry made the offer that if the Swansea City Council were to build an astronomical observatory dome on the top of the Observation Tower, the Swansea Astronomical Society would equip it and make it possible for the public and astronomers alike to enjoy astronomical experiences throughout the year.

His initial suggestions were that a Camera Obscura type device could be used to display terrestrial views and images of the Sun and Moon. Gerry's astronomical hero was Norman Lockyer and his main interest was the Sun, so it is not surprising that his initial suggestions were that a device for safely observing the Sun could be installed in the Tower and that members of the public using the promenade would be able to view the images produced. On 4 November 1984, following this meeting,

Gerry Lacey wrote to the Local Authority outlining his proposals on behalf of the Society.

It must have been a devastating blow to Gerry when he received a reply to his letter from the Chief Executive and Town Clerk of Swansea on 19 December 1984. He was told that the City Officers had given consideration to his proposal and that they did not feel that it merited further discussion at this stage. However, the door was left open and the Chief Executive suggested that prior to the budget for 1986/87, a more comprehensive report on the proposal could be formulated.

It would have been very easy to give up at this stage, but on 20 January 1986 Gerry Lacey again wrote to the Chief Executive, pointing out that a recent series of open evenings held at the Swansea Astronomical Society's Fairwood Common Observatory site confirmed public interest in astronomy and suggested that a suitably located public observatory could enhance tourism in the area. One thing was certain. Gerry Lacey had made it plain that a building, which could have been regarded as a folly, funded partly by European and partly by local money, had the potential to be something far more exciting and useful. The seed was sown.

An Observatory Needs a Telescope

Of course, letters went back and forth between Gerry Lacey and the Local Authority, and on 12 April 1987 Gerry again wrote to the City Chief Executive enquiring about the proposed project. On 15 April, Gerry received a letter from the Chief Executive, saying that provision had been made in the City Council's Capital Budget for 1987/88 for a start on the project with a view to completion in 1988/89. The Swansea Astronomical Society was asked to formally confirm its interest in the use of the Tower. The game was on.

An observatory is not just a building. It also comprises the equipment that allows astronomical observation to take place. Local government was going to deliver their part of the deal: An observatory tower was to be built. The Astronomical Society also had to deliver on a scale commensurate with this. This is where the late Dr. Fred Jenkins enters the story.

Fred was a retired medical general practitioner and anaesthetist. He was also a capable mechanical engineer and held several patents for mechanical medical devices. He had been building astronomical telescopes since the 1950s and had been a member of the Swansea Astronomical Society for many years. Fred offered to build a unique telescope for the new observatory. He offered to construct a 0.5 m Shafer–Maksutov.

In 1944 the Russian Academician Dr. D.D. Maksutov published an article in the *Journal of the Optical Society of America*. He explained that by putting a negative corrector lens at the front of the scope, it is possible to use a spherical primary mirror to achieve the same optical effect as parabolizing the primary mirror, which brings the image to one sharp focus and eliminates spherical aberration. As it is easier to make a spherical mirror and the negative corrector lens than it is to subsequently parabolize the mirror, it should be easier to make a Maksutov telescope than a telescope requiring a parabolized mirror. Maksutov had also pointed out that it would be possible to make the corrector smaller by placing it further down the light path in the converging beam of light. Strangely, nobody seemed to take note of this last point, although Maksutov's basic telescope design was widely adopted. Maksutov–Cassegrain telescopes are now commonplace and are available from the major telescope manufacturers around the world. However, they are usually of a classical Maksutov design, with the corrector lens at the front of the telescope tube.

In 1980, almost 40 years after Maksutov suggested placing the corrector optics further down the light path, David Shafer, the American optics designer, published a design for a modified Maksutov that incorporated this feature. Fred Jenkins decided to build a 0.5 m version of this Shafer – Maksutov. Fred made the corrector optics himself. They comprised a concave and a convex lens separated by air, and placed in front of a convex zerodur secondary mirror. The light thus passes through the corrector optics twice on its way from the mirror to the eyepiece. The glass manufacturer Corning UK cast the 0.5 m primary mirror blank. They had never previously made an optically annealed large glass disk in Britain before and had to send someone to Corning USA who had experience in manufacturing large mirrors. David Sinden, the British optical engineer, finished the 0.5 m spherical mirror. Fred had obtained the help and cooperation of numerous individuals and local companies to help him complete the telescope. The scope was finished by 1990 and Fred had the frustration of waiting while the building caught up.

The Observatory Buildings

The Marina Towers Observatory is actually two buildings (Figure 18.1). The seaward, southern building is square in cross-section and has five floors. The top four floors stand on four corner pillars, which extend down through the larger ground-floor exhibition room to the foundations. The landward, northern building is circular in cross-section. This building houses an iron spiral staircase. Doors open out onto bridges, which join the two buildings at each floor. The staircase being housed in a separate building means that the Observatory is extremely safe from fire. Should fire break out on any floor, people on higher floors only have to cross the outdoor bridge between the Observatory building and the staircase tower and descend to safety. The staircase, being located in a separate building, causes minimal vibration to be transferred to the Observatory building. The apparent joining of the two buildings by bridges is only an illusion, for the bridges are attached to the square building and are actually separated from the staircase building by a small gap (Figure 18.3).

The staircase building houses a model of the Solar System held on a wire in the central stair-well (Figure 18.4). Each step on the staircase represents a distance of 46 million miles on the journey from the Sun's disk at ground level to diminutive Pluto at the entrance to the Observatory dome at the top of the building, which is capped by a magnificent stained glass roof (Figure 18.5). Information on each of the planets is presented on wall plaques which may be read as one ascends the staircase.

The Dome is a unique design by Norman Walker who spent a number of years at Herstmonceux before becoming a dome-maker. The dome is 15 feet in diameter and actually floats on a saline water bearing (Figures 18.6 and 18.7).

Salt is added to the water bearing to provide extra buoyancy and to prevent the water from freezing in cold weather. A thin layer of oil on the surface of the saline water helps reduce evaporation. However, evaporation does take place and high winds can blow water out of the trough. This means that the dome moves slightly up and down as the water level changes. The dome is prevented from lifting out of the trough,

Figure 18.3. View from beneath a bridge showing the gap between the bridge and the staircase building.

Figure 18.4. The Solar System model was built by local schoolchildren.

Figure 18.5. The stained glass roof is visible at the top of the staircase. Ringed Uranus and distant Neptune are also visible.

and is held centrally within the trough, by five pairs of wheels mounted equidistantly on the inside of the bearing trough (Figure 18.8). The wheels with the vertical axles hold the dome's circular float centrally

Figure 18.6. The dome is like a circular boat that floats on its water bearing. Here is the dome before it was hoisted into position (photograph by Mavis Morgan).

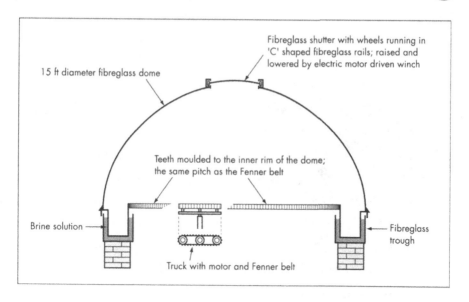

Fibreglass shutter with wheels running in 'C' shaped fibreglass rails; raised and lowered by electric motor driven winch

15 ft diameter fibreglass dome

Teeth moulded to the inner rim of the dome; the same pitch as the Fenner belt

Brine solution

Fibreglass trough

Truck with motor and Fenner belt

Figure 18.7. The dome construction showing the fibreglass dome floating on a bearing of saline water in the circular trough (drawing by Fred Whittle).

within the trough and the wheels with the horizontal axles prevent the dome from lifting out of the trough.

On mid-summers day in 1991 a crane lowered the dome into place (Figure 18.9).

In June 1992, the Swansea Astronomical Society started the long work on the interior of the Observatory and the installation of the telescope. The Marina Towers Observatory was officially opened on 24 September 1993 and the building was named "The Tower of the Ecliptic".

Figure 18.8. Wheels mounted at right angles to each other on the inner rim of the dome-bearing trough serve to hold the dome in place while allowing free rotation and a little vertical movement.

Figure 18.9. The dome is lowered into place by crane (photograph by Mavis Morgan).

The large, ground floor exhibition room has wide, arching windows on three sides and houses the Gerry Lacey Exhibition, comprising astronomical instruments and the results of work done by members of the Society (Figure 18.10). Every Wednesday morning throughout the school year, schoolchildren enjoy the astronomical activities, talks and a visit to the dome to see the telescope (Figure 18.11). Some astronomy classes taught by Society members for the Department of Adult, Continuing Education, University of Wales, Swansea, are held in the Observatory during the evening.

The first floor houses a photographic darkroom, which is also used for slide-shows on open nights (star parties). The library (Figure 18.12) is located on the second floor. I am writing this chapter in the library, which houses a fine collection of astronomical texts and overlooks the bay. Small meetings and discussion groups make use of this facility in addition to its main library function. The third

Figure 18.10. Summer visitors look at the exhibition and talk with Society members.

Figure 18.11. Schoolchildren explore the Earth–Moon distance using models in the exhibition room.

Figure 18.12. A Society member reads an astronomical text in the tranquillity of the library.

floor comprises an observation platform, which is also used by a sailing club for signalling.

During the summer months, the Marina Towers is opened to the public on two afternoons of the week (Figure 18.10) and tourists have the opportunity to see Swansea's unique Observatory and telescope. On sunny days, images of the Sun, captured by a video-camera mounted on a small solar telescope in the Observatory dome, are transmitted from the dome to monitors in the exhibition room. The solar eclipse of August 1999 (which was 97% in Swansea) was viewed in safety by hundreds of visitors to the Observatory. For seven months of the year from September to the end of March, monthly night-time star-parties are held for the public when they have the opportunity to view the Moon and planets for themselves. For people unable to climb the staircase, live video images of the Moon and planets are transmitted to the exhibition room at street level.

The Marina Towers Observatory is a focal point on the promenade of a forward-looking Welsh city. It is an educational and tourism resource, providing part of Swansea's unique character. It is an interface between an active, outgoing astronomical society, schools and the public. Being sited in a light-polluted area, the Observatory demonstrates that worthwhile astronomical observations of the Solar System can be made even under adverse conditions. Most of all, it shows what can be achieved when Local Government and an astronomical society get together to reach for the stars.

Contributors

Chris Anderson lives in Kentucky, USA. He has been a nature and astrophotographer since the early 1980s, concentrating on what he refers to as "backyard" nature photography. His interest in photography actually began in the late 1970s when he developed a passion for astro-imaging. Today, he maintains Darklight Observatory, dedicated to digital imagery of the cosmos.

A Further and Higher Education lecturer in Photography and Multimedia, **Paul Andrew** has been interested in astronomy since the age of eleven. He formed the South East Kent Astronomical Society in 1972 and holds the position of Honorary President. Very much a visual observer, Paul has a strong interest in deep-sky objects, preferring the observation of more elusive objects to that of the "classical" favourites.

By training and profession, **Bill Arnett** is a software engineer based in California, USA. He says that over the years he has tried to learn a little about a lot of things, and sometimes envies the great men of the Renaissance and the Enlightenment who had a very real possibility of learning a large fraction of all human knowledge! But reflecting on what mankind has achieved since then, he admits that he has the better deal.

Tom Boles is a retired computer engineer and former telescope maker. He is a Fellow of the Royal Astronomical Society and is on the Council of The British Astronomical Association.

Michael Brown lives in York, England. He developed a keen interest in astronomy while still at school, later he started making and using telescopes. In 1998 he entered the digital world of CCD imaging of the Moon and planets and says he looks forward to many more years of what has been a most satisfying hobby.

Ken Dauzat is from Louisiana, USA. He writes, "Hi, I am 51 years old and have a lovely wife, Karen, two beautiful daughters, Angel and Ashly, and a fine son, Dwayne. We live in the country, in central Louisiana, about 20 miles south of Alexandria, at a small area called Hickory Hill. I run an Internet-based custom machining business that builds telescope rings and accessories. Astronomy is part of my life and I love observing and helping others in the hobby. I hope my observatory and this book encourages others to do the same."

Paul Gitto is a retired dentist living in Whiting, New Jersey, USA, with his wife Anne, and their three sons. He is currently upgrading his observatory for near-Earth asteroid research. He is the Webmaster for the local astronomy club, ASTRA. You can e-mail him at cometman@cometman.com and his Web site is at http://cometman.com

Alan Heath is a retired schoolteacher. He served as Director of the Saturn section of the British Astronomical Association for thirty years, and contributes to the work of the BAA and also to the Association of Lunar and Planetary Observers in the USA.

Rob Johnson is a technical manager for a photoengraving plate manufacturer. He has been an active member of the Liverpool Astronomical Society for over 30 years and has a keen interest in astrophotography and CCD imaging. Rob can be contacted by email at robjn@blueyonder.co.uk.

George Kolovos is a retired electronics engineer. He lives in Thessaloniki, Greece. He has been an amateur astronomer since he was 15 years old. From 1966 until 1997 he worked as a technician in the Astronomical Laboratory in the Observatory of the Aristotle University of Thessaloniki, Greece. He is interested in a big variety of celestial objects, such as the Moon, the Planets, Variable Stars and especially the Sun, which is the only object he has been observing for the last few years.

Martin Mobberley served as British Astronomical Association President from 1997 to 1999 and received the BAA's Goodacre award in 2000. For over twenty years he has been an active astrophotographer and imager of comets, planets, novae and supernovae. By day he is a Marconi software engineer, writing C and assembly language routines for microprocessors. He is the author of *Astronomical Equipment for Amateurs,* published in this series.

Alf Jacob Nilsen was educated as a biologist and works full time as a teacher in the secondary school at Flekkefjord, a small town in the south of Norway. In addition to studying tropical marine biology, his hobbies are photography and astronomy. He is a member of the Norwegian Astronomical Society.

Peter Paice was born in 1932 in Eastbourne, Sussex, UK. On retirement from teaching at the Methodist College, Belfast, he has expanded his lifetime hobbies of electronics and photography, including astrophotography. An enthusiast inventor, he has combined practical skills and lateral thinking to construct remotely controlled telescopes and CCD devices. A convert to digital cameras, he has developed afocal coupling to telescopes. Because of constraints imposed by the observing site and climate, specialisation has been in solar and planetary imaging. His images have been widely published in professional journals, and on websites. You can email him at papaice@aol.com and his Website is at http://www.spaice.co.uk.

David Ratledge lives in Lancashire, UK. He started building telescopes over 40 years ago and is still dreaming of building bigger ones today! He has written or edited three books: *The Art and Science of CCD Astronomy, Software and Data for Practical Astronomers,* and *Observing the Caldwell Objects* – all published by Springer. He is a regular contributor to *Sky & Telescope* magazine for which he is an Associate Editor. His Web site is www.deep-sky.co.uk.

Gordon Rogers's interest in astronomy took off on 13 June 1994 when he saw through a small refractor what he considers was probably the effect of three asteroids hitting the moon. He did not then have the means to image celestial objects. Now he has the means "big time". Nothing has happened on the Moon and his interest has turned to deep space. He was privileged to have Sir Patrick Moore film an edition of BBC TV's *The Sky at Night* at his observatory at Crendon in England.

Bob Turner is a lecturer in Astronomy and Astrophysics and is a regular speaker at many of the Astronomical Societies in the South of England. Specializing in solar research, Bob has over the years done much work in hydrogen-alpha, looking at periodicity in solar flares and pre- and post-flare stages. Bob has a number of solar telescopes at his observatory in West Sussex, ranging from a 6 $\frac{1}{2}$-inch refractor to a portable 3-inch that goes in his car.

Dr Stephen Wainwright is a Senior Lecturer in Biological Sciences at the University of Wales, Swansea. A Fellow of the Royal Astronomical Society and a member of the Cardiff and Swansea Astronomical Societies, he is Webmaster for the Swansea Astronomical Society and also the founder of the QCUIAG international astronomical imaging group which currently has more than 2000 members worldwide. It's on www.qcuiag.co.uk.

Paul Zelichowski is a 44 year old Canadian from Ontario. He has been interested in astronomy (and science in general) since he was a young lad of eight or nine. He had to wait until he was 23 before he obtained his first telescope, a 6-inch reflector. These days he images mostly from the comfort of his dining room, using several computers, CCD cameras, and a remote-controlled mounted 10-inch reflector. He says, "It's basically a dream come true for me."

About the CD-ROM

The CD-ROM contains a single file called **Small Observatories 1.pdf** and a directory called **Acrobat.**

You need Acrobat Reader installed on your computer to use this CD-ROM. It is quite likely that you already have it – in which case all you need to do to view the first volume, *Small Astronomical Observatories, edited by Patrick Moore*, is to put the CD-ROM into your computer and, using Windows Explorer, double-click on the file.

Alternatively, you can open the file **D:\Small Observatories 1.pdf** (assuming your CD drive is drive D:) from within Acrobat Reader.

Once you have opened the book, you can view it or print it out. You can use the index at the front of the book to locate chapters via Acrobat, but you should add 9 to the page number shown, to account for the preliminary pages.

Small Astronomical Observatories, edited by Patrick Moore, is © Springer Verlag London Limited 1998.

You may print out the materials on this CD-ROM for your own use only. This material is copyright and you may not print multiple copies of any or all of it, copy the electronic files, sell, hire or otherwise distribute the material or any part of it.

Installing Adobe Acrobat Reader

Two versions of Acrobat Reader are available on the CD-ROM. They are both located in a directory called **Acrobat**. The 32-bit version for Windows 95 or higher is in the sub-directory **\32bit**. If you are still using Windows 3.1, install the version in the sub-directory **\16bit**. Simply go to the relevant directory in Windows Explorer (or File Manager in 3.1) and double-click on the ''.exe'' file. That will install Acrobat Reader.

Make sure there are no other applications running when you install Acrobat Reader. Acrobat Reader takes up only about 5MB of disk space.

Adobe, Acrobat and their logos are trademarks of Adobe Systems Incorporated. Windows is a registered trademark of Microsoft.